MW01142780

Reflections on Modern Psychiatry

Reflections on Modern Psychiatry

Edited by

David J. Kupfer, M.D.
Professor and Chairman
Department of Psychiatry
University of Pittsburgh School of Medicine
and
Director of Research
Western Psychiatric Institute and Clinic
Pittsburgh, Pennsylvania

Washington, DC
London, England

Note: The authors have worked to ensure that all information in this book concerning drug dosages, schedules, and routes of administration is accurate as of the time of publication and consistent with standards set by the U.S. Food and Drug Administration and the general medical community. As medical research and practice advance, however, therapeutic standards may change. For this reason and because human and mechanical errors sometimes occur, we recommend that readers follow the advice of a physician who is directly involved in their care or the care of a member of their family.

Books published by the American Psychiatric Press, Inc., represent the views and opinions of the individual authors and do not necessarily represent the policies and opinions of the Press or the American Psychiatric Association.

Manufactured in the United States of America on acid-free paper
95 94 93 92 4 3 2 1

American Psychiatric Press, Inc.
1400 K Street, N.W., Washington, DC 20005

Library of Congress Cataloging-in-Publication Data
Reflections on modern psychiatry / [edited by] David J. Kupfer.
p. cm.
Based on the International Symposium on Clinical Research in Affective Disorders and Schizophrenia, held in Pittsburgh, Pa., Apr. 22–23, 1989.
Includes bibliographical references.
ISBN 0-88048-516-7
1. Biological psychiatry—Congresses. 2. Affective disorders—Congresses. 3. Schizophrenia—Congresses. 4. Psychiatry—Congresses. I. Kupfer, David J., 1941– . II. International Symposium on Clinical Research in Affective Disorders and Schizophrenia (1989 : Pittsburgh, Pa.)
[DNLM: 1. Mental Disorders—congresses. 2. Psychiatry—trends—congresses. WM 100 D489 1989]
RC327.D49 1992
616.89—dc20
DNLM/DLC
for Library of Congress 91-33208
CIP

British Library Cataloguing in Publication Data
A CIP record is available from the British Library.

Table of Contents

Contributors

Donald W. Black, M.D.
Associate Professor of Psychiatry, University of Iowa College of Medicine, Iowa City, Iowa

Floyd E. Bloom, M.D.
Chairman, Department of Neuropharmacology, The Scripps Research Institute, La Jolla, California

Joseph V. Brady, Ph.D.
Professor of Behavioral Biology, Department of Psychiatry and Behavioral Sciences and Professor of Neuroscience, The Johns Hopkins University School of Medicine, Baltimore, Maryland

Arvid Carlsson, M.D.
Emeritus Professor, Pharmacology Institute, Göteborg University, Göteborg, Sweden

Thomas Detre, M.D.
Distinguished Service Professor of Health Sciences and Professor of Psychiatry, University of Pittsburgh, Pittsburgh, Pennsylvania

Daniel X. Freedman, M.D.
Judson Braun Professor of Psychiatry and Pharmacology and Executive Vice Chairman, Department of Psychiatry and Biobehavioral Sciences, University of California, Los Angeles School of Medicine, Los Angeles, California

Gary A. Gudelsky, Ph.D.
Associate Professor of Psychiatry and Pharmacology, Case Western Reserve University School of Medicine, Cleveland, Ohio

Samuel B. Guze, M.D.
Spencer T. Olin Professor, Department of Psychiatry, Washington University School of Medicine, St. Louis, Missouri

Philip S. Holzman, Ph.D.
Esther and Sidney R. Rabb Professor of Psychology, Department of Psychology, Harvard University, Cambridge, Massachusetts; Director, Psychology Laboratory, Mailman Research Center, McLean Hospital, Belmont, Massachusetts

David J. Kupfer, M.D.
Professor and Chairman, Department of Psychiatry, University of Pittsburgh School of Medicine; Director of Research, Western Psychiatric Institute and Clinic, Pittsburgh, Pennsylvania

Herbert Y. Meltzer, M.D.
Douglas Bond Professor of Psychiatry, Case Western Reserve University School of Medicine, Cleveland, Ohio

Christopher Mountjoy, M.B.B.S., F.R.C.P(sychiatr)
Department of Psychiatry, Addenbrookes Hospital, University of Cambridge, Cambridge, England

Amelia Nasrallah, M.S.
Department of Psychiatry, University of Iowa College of Medicine, Iowa City, Iowa

Sir Martin Roth, M.D., F.R.C.P(sychiatr), Sc.D.(Hon)
Emeritus Professor, University of Cambridge; Department of Psychiatry, Addenbrookes Hospital, and Fellow of Trinity College, Cambridge, England

Gary J. Tucker, M.D.
Professor and Chairman, Department of Psychiatry and Behavioral Sciences, University of Washington School of Medicine, Seattle, Washington

George Winokur, M.D.
Paul W. Penningroth Professor, Department of Psychiatry, University of Iowa College of Medicine, Iowa City, Iowa

Thomas Detre, M.D., is not only a psychiatrist but one of the nation's most accomplished academic medical administrators. Throughout his career, he has demonstrated keen academic judgment and a strong "vision" for the future of psychiatry and medicine as well as the health sciences. Dr. Detre is currently Senior Vice President for Health Sciences of the University of Pittsburgh and President of the University of Pittsburgh Medical Center. He was formerly President of the Medical and Health Care Division (1986–1990), Associate Senior Vice Chancellor for Health Sciences (1982–1984), and Chairman of the Department of Psychiatry at the University of Pittsburgh School of Medicine (1973–1982). Since 1973, he has also served as Director of Western Psychiatric Institute and Clinic. He came to Pittsburgh from Yale University, where he was Professor of Psychiatry and Psychiatrist-in-Chief of Yale-New Haven Hospital.

Dr. Detre was born in Budapest, Hungary, where he began his studies of psychiatry and medicine. After completing his medical education at the University of Rome in 1952, Dr. Detre came to the United States and served an internship at Morrisania City Hospital in New York City (1953–1954). He received his postgraduate training in psychiatry at New York's Mount Sinai Hospital (1954–1955) and at Yale University (1955–1958). He was first appointed to the faculty of Yale's Department of Psychiatry in 1957 and served there for the next 15 years.

Dr. Detre's interests are wide-ranging, and he has published numerous articles and chapters on recurrent depression, violence and aggression in children, and other biologic aspects of mental disorders as well as on health policy issues. He has also played a leadership role in national activities on health and research policy development, serving as chairman of various study sessions and advisory committees of the National Institutes of Health, the Alcohol, Drug Abuse and Mental Health Administration, the Food and Drug Administration (Psychopharmacologic Drugs Advisory Committee, 1984–1986), and the U.S. Department of Veterans Affairs (Secretary's Advisory Committee for Health Research Policy, 1990–1991).

Foreword

The Future of Psychiatry—as Seen From Its Past

I owe my colleagues a warm expression of gratitude for preparing the series of fine, thoughtful papers that were originally presented at a symposium in my honor and now appear in this volume. These chapters, which review the past three decades of progress in psychiatry and attempt to foresee its future, offer clear evidence that the profession today is not the one whose ranks I joined 30 years ago. And, if we have learned anything from the lessons of hindsight, the psychiatry of the 21st century—if it is even called psychiatry—will, of necessity, evolve into a bold new hybrid that pulls together the basic, the clinical, and the social sciences, avoiding the unidimensional views of mental illness that have misled us in the past.

When I entered psychiatry, the specialty relied far more heavily on psychology and the social sciences than on neurology and other neurosciences as the most likely source of further understanding of mental illness. The profession's decision to disaffiliate from neurology in the late 1950s and early 1960s—a decision with which I did not concur, by the way—set a historical precedent: psychiatry became the first medical specialty without ties to any organ or organ system.

Fortunately, as demonstrated in the chapters that follow, some colleagues shared my bias and vigorously sought to prove its merit. Evidence of the biological and genetic bases of disorders such as depression and schizophrenia mounted steadily and incontrovertibly. Psychiatry reworked diagnostic systems that had been based more on majority vote than scientific data. The emergence of effective medications for mental disorders, particularly the antipsychotics and antidepressants, gave us new therapeutic alternatives that actually seemed like real medicine, both to our patients and to us. In short, after embracing a world view that was in perfect tune with the spirit of the early 1960s, we reversed ourselves and attempted a rapprochement with the medical mainstream. We began to encourage our residents to seek training in neurology; we taught psychopharmacology with the enthusiasm that had been reserved for psychotherapy supervision; and we attempted to reembrace populations such as elderly patients, patients with dementia, and patients who were developmentally disabled—individuals we had long ago relinquished to the ministrations of psychologists and special educators. What is astounding, however, is how little effort we made to test the validity of our theories.

This foreword is adapted from an article, "The Future of Psychiatry," which appeared in the May 1987 issue of the *American Journal of Psychiatry*.

Although we tried to justify the unsound reasoning underlying many of our assumptions about mental illness on the basis of the alleged mind-body dualism, nothing could be farther from the truth. The source of the confusion is that although the "mind" and mental functions constitute a legitimate and convenient conceptual framework to describe certain phenomena about psychiatric illness, they are not avenues for the generation of biological theories by which we can deduce from the nature of mental defect its etiology and pathogenesis. The history of medicine is replete with examples demonstrating that our failure to detect changes in the functioning or structure of organs or organ systems proves only the limitations of our diagnostic methodology and cannot be construed as evidence for the psychological origin of etiology or pathogenesis.

We do not have to feel lonely in our disarray, however: our half sibling and our stepparent—psychology and neurology, respectively—have also been struggling with their own identity crises. In psychology, the result has been the fragmentation of the discipline into a number of oft-warring schools. Practitioners of neurology and psychiatry have both engaged in various more or less successful Scotch-taping measures to strengthen their scientific underpinnings. Although speciality training in psychiatry and neurology now includes some exposure to clinical pharmacology, neither discipline provides sufficient knowledge of how the pathophysiology and treatment of various disorders affect the functioning of the nervous system. Recent advances in neurochemistry, neuroendocrinology, and computer-assisted electrophysiologic and imaging methods, for example, should have acted as a major impetus for the investigation of various encephalopathies and the attendant changes in mental functioning, but this is not happening.

Even so, certain disorders will continue to manifest themselves not with observable neuroendocrine or neuroanatomical changes but in subtle cognitive and personality alterations. We cannot molecularize mood and cognition. In addition, certain synthetic neural functions are expressed in a psychological mode. Thus, in embracing the promise of neuroscience, we cannot throw away psychology. Ignoring psychological functioning would not advance the field.

The real problem, and we might as well face up to it, is that the body of knowledge we need to further our understanding of psychiatric disorders no longer fits comfortably within the self-imposed boundaries of our various specialty disciplines. In my opinion, the only viable solution is a new career path leading to specialization in what I would call (for lack of a better term) clinical neuroscience. It should be concerned with disorders of the nervous system, regardless of whether the etiology of the disorder is known or unknown or whether the disorder is primary or secondary. Trainees should have adequate exposure to medicine, neurology, psychiatry, and pharmacology and should also be taught what neuropsychology, electrophysiology, and imaging methods can contribute to the assessment of brain functioning in living patients. Although training in neurosciences is important, acquisition of basic skills in epidemiology and biostatistics is no less significant. We must admit that the time for "Lone Ranger" physicians who know everything about everything has long passed. Subspecialization in an area of choice—be it developmental disorders, metabolic encephalopathies, aging, neuromuscular disorders, so-called functional psychoses, or whatever—should follow an integrated and carefully thought out curriculum in medicine, neurology, and psychiatry, which could take approximately 5 years of study.

Research training should follow clinical training and be tailored to the trainee's needs. Regardless of whether the medical students' career choices will involve patient care, clinical investigation, or laboratory science, the emphasis in their training should be on the critical examination of problem solving and validation methods to both prepare young physicians to handle the explosion of new information and to protect them from the embarrassing mistakes our generation has made. Clinical skills should be taught by active investigators, who by doing research have a clear appreciation that today's facts are merely way stations on a journey to greater knowledge—facts that should be taken seriously only until new findings have modified them or rendered them valueless.

Failure to achieve what one might broadly call psychosocial competence in our trainees, however, would merely create another kind of reductionism, replacing the brainless with the mindless approach to clinical problems. What Eisenberg (1979) calls the "sociobiological process of becoming ill, being ill, and getting well," what social anthropology can teach us about patienthood and illness, and what political science contributes toward the understanding of health care delivery systems should be learned by all physicians (Waitzkin 1984). Nor do I not believe that trainees in any one specialty should bear sole responsibility for mastering interviewing techniques or psychotherapy. Every physician must learn how to gather data that are critical for diagnostic and treatment decisions and to do so in a way that is vigorous and precise without being relentless or insensitive. Last but not least, all physicians, whether they are engaged in research or clinical care, should be trained to practice the informed consent doctrine. Discussing with our patients the pros and cons of various diagnostic and treatment options is not just a moral obligation but a major contribution to our own education. The very process makes it mandatory for us to acknowledge how limited our information is—and encourages us to behave modestly.

I might as well admit that my views find little support in psychiatric circles, despite the fact that I have made something of a career of being a reasonably accurate predictor of psychiatry's future. (Some have even given me credit for helping to assure that my predictions were borne out). I end, then, where I began, preferring to fall back on the comfort of my own predictions than to trust those of someone else. As we learn from ever more persuasive evidence that the mind and the brain are one and the same, we must—as practitioners of a specialty intent on evolving in a way that assures survival—abandon the false boundaries we have established in our game of king-of-the-mountain with other medical and health care disciplines. Our new strategy should be a decisive one, carefully conceived, meticulously executed, and dedicated to providing clinical and research training to create a new breed of psychiatrists who are truly neuroscientists. This, I believe, is psychiatry's future.

Thomas Detre, M.D.

References

Eisenberg L: Interfaces between medicine and psychiatry. Compr Psychiatry 20:1–10, 1979

Waitzkin H: Doctor-patient communication: clinical implications of social scientific research. JAMA 252:2441–2446, 1984

Preface

The genesis of this book is rooted in a desire to honor a man who has distinguished himself through his important contributions to clinical research in psychiatry. Throughout a career spanning more than 30 years, Thomas Detre has displayed exceptional acumen as both a psychiatrist and an academic administrator. His ability to conceptualize and facilitate multidisciplinary, collaborative research programs has strongly influenced contemporary research in psychiatry. Dr. Detre's vision of the future of psychiatry and its potential to contribute to greater understanding of other medical illnesses has been incorporated into a model for academic medicine that has been adopted by academic medical centers throughout this country and abroad.

Several years ago, we began to think about an appropriate way to honor Dr. Detre and his accomplishments. After contemplating several options, we decided that the most meaningful gesture would be the organization of an international symposium. Dr. Detre's long-standing interest in major psychiatric disorders and issues related to the medical management of these mental health problems prompted us to arrange a meeting that addressed contemporary diagnostic and treatment practices. As it evolved, we realized that the theme selected for our meeting covered topics explored earlier by Dr. Detre and Dr. Henry Jarecki in their 1971 book *Modern Psychiatric Treatment*. For the symposium in honor of Dr. Detre, we invited distinguished scientists and clinicians to review the advances in our understanding of affective disorders and schizophrenia over the past 25 years and to offer their thoughts about the future of the field. The following pages contain the perspectives shared by these eminent scientists at the International Symposium on Clinical Research in Affective Disorders and Schizophrenia, which was held in Pittsburgh, Pennsylvania, on April 22 and 23, 1989. The multidisciplinary content of the papers, the adherence to rigorous scientific standards, and the willingness to go out on a limb with ideas and opinions reflect several of the qualities that characterize Dr. Detre's approach to clinical research in psychiatry. This group of highly respected researchers and clinicians have prepared chapters that embody many of the elements of scientific excellence advocated by Dr. Detre. We believe that the resulting collection of papers is a fitting tribute to a man whose accomplishments are so worthy of recognition.

David J. Kupfer, M.D.

References

Detre TP, Jarecki HG: Modern Psychiatric Treatment. Philadelphia, PA, JB Lippincott, 1971

1

Diagnosis in Psychiatry: Philosophical and Conceptual Issues

Samuel B. Guze, M.D.

The philosophy of diagnosis is concerned with the theory of diagnosis: the assumptions, strategies, and implications of diagnosis, as well as the constraints affecting it. It is important for us to try to understand these matters, because our understanding will shape our expectations and approaches to diagnosis.

Medical diagnosis is primarily an effort to classify individuals with regard to their health status. It is used principally to classify patients (those who come or are referred because of some manifestation of, or concern about, illness) or to classify individuals as part of a health survey (whether as part of regular checkups or of epidemiological studies). The basic assumption underlying diagnosis is that, even though every individual and every individual experience of illness is unique, individuals can be grouped in meaningful and important ways to provide useful and often essential insights about their illnesses.

Depending on the state of knowledge at any given time, assigning a diagnosis (within the limits of diagnostic error) may convey information concerning etiology, pathogenesis, appropriate treatment, course and outcome, familial illness patterns, overall costs of illness and treatment, and much more. A diagnosis may also convey uncertainty about the nature of the condition and point the direction for further study of the individual patient (Guze 1984.)

Although the diagnostician may be interested only in his or her own clinical experience, it is obvious that diagnosis is embedded in a social context and is indispensable for communication and comparison. A diagnosis is necessary if we are to estimate how closely our experience matches that of colleagues, whether they be down the hall or around the world. Everyone does not have to use the same diagnostic labels or categories, but we must make the process of how we make our diagnoses as clear and operational as possible and offer enough information to permit others to use our experience to clarify their diagnoses.

This is not always possible, of course. Not all diagnosticians agree on the parameters or criteria for assigning diagnoses. There may be major differences in the way patients are observed and studied so that very different data are obtained, making some comparisons impossible. Systematic studies are required to resolve such an impasse by establishing the validity of any particular diagnostic category: to show that the assigned diagnosis is differentially associated with other important variables, such as etiology, pathogenesis, and response to treatment. If different diagnostic systems achieve similar validity across such variables, some diagnosticians will feel constrained to understand and explain such findings. They will devise and carry out studies in which patients are classified simultaneously according to more than one system, looking for insights that might enlarge our understanding. *Validation is the gold standard for evaluating any diagnostic system and for comparing one system to another.* Aesthetic and political considerations and a priori assumptions must give way to validity if there is to be progress in this field.

The most important implication inherent in the diagnostic effort is the assumption or belief that it is possible and important to classify illnesses. However, such classification makes sense only if one believes that there are important differences correlated with different diagnoses. This means, of course, that one accepts the principle that, associated with any particular diagnosis, there are common denominators in the disorder or illness that transcend individual variations or differences. Thus, one accepts that the diagnosis of cirrhosis of the liver, epilepsy, diabetes, or bipolar disorder—as examples—tells us something important for our approach to the patient that is independent of the individual's age, sex, race, education, marital status, sociocultural context, or other demographic variables.

Diagnostic categories and criteria, however, are subject to change as new ideas and new findings become available. Scientific progress inevitably leads to new thinking about diagnosis. Traditional or long-used criteria may be qualified or even replaced by entirely new criteria. The latter may reflect new parameters, such as neuroimaging findings, biopsy results, or measurements of biochemical or physiological processes. Sometimes clinical symptoms and signs will take on differential significance depending on the presence or absence of a particular laboratory finding.

For example, it may make at least heuristic sense for research purposes to classify patients with major depression into those with a positive or negative dexamethasone suppression test or those with or without specific sleep patterns. If such distinction leads to significant differences in response to treatment, long-term outcome, or familial transmission patterns, the distinction will be seen as validated and will become generally accepted. On the other hand, if molecular geneticists can identify a gene or genes responsible for bipolar disorder, the presence or absence of the gene may become the principal criterion for diagnosis, and the presence or absence of symptoms and signs may be seen as modifying criteria. Under such circumstances, it may turn out that some individuals carrying the gene(s) will show only the classical mood and energy features, whereas others will show equally prominent delusions and hallucinations. Furthermore, other individuals may show unexpected clinical features not at all like typical bipolar patients, and still others may never show any of the recognized clinical features. Subclassifying individuals into such groups may turn out to be associated with important differences in one or more significant parameters of study, thus justifying the differentiation and leading to major new insights about etiology or pathogenesis.

Although diagnosis is clearly an indispensable component in efforts to study, treat, and prevent illness, it is not an end in itself. Diagnosis is limited by our knowledge. If we know little about the etiology or pathogenesis of any disorder, it is highly likely that our approach to the diagnosis of the disorder will be greatly constrained. It follows that we must do everything we can to improve our present diagnostic system(s), even as we recognize that considerable progress has taken place in recent years. We need at the very least to reduce ambiguities further and improve the definition of individual criteria. At the same time, we must recognize that significant improvement can come only through systematic, controlled studies designed to assess the validity of the diagnostic category.

As I have already emphasized, validity is the key to improved diagnostic criteria and the touchstone for progress. Differential validity is the scientific way to determine the relative power and worth of competing diagnostic systems or criteria. The goal is to achieve the highest accuracy in the predictions inherent in diagnosis. Diagnostic categories should, ideally and ultimately, specify etiology, pathogenesis, and response to various treatments as well as genetic and epidemiologic attributes. We are a long way from achieving this goal for any medical diagnosis, but specifying a goal helps define the strategies for achieving it. It is conceivable that before too long—for some psychiatric illnesses at least, as well as for some general medical disorders—our diagnostic categories will specify clinical signs and symptoms, physiological deviations, methods of ascertaining the case, relevant genotype, personality, socioeconomic status, and associated illnesses.

One of the issues concerning diagnostic validity is whether we are talking about "real" entities or only "apparent" ones when we assign individuals to diagnostic categories. This issue may be settled only when we can use our categories to define specific and different etiologies and/or pathogenetic processes. Until such a time, one's position will reflect one's strategic thinking about psychiatric disorders. If one believes, as many of us do, that schizophrenia is as "real" as heart failure, then one tries to define and study schizophrenia as one defines and studies heart failure: clinical presentations, modes of onset, age of onset, course, prognosis, outcome, complications, response to interventions, physiological deviations, and other relevant criteria. One accepts the possibility, even the likelihood, that the syndrome called schizophrenia may include cases representing different etiologic agents or pathogenetic mechanisms, just as is true with heart failure.

Another important concern of psychiatric diagnosis is that the current syndromatic approach, as exemplified in DSM-III or DSM-III-R (American Psychiatric Association 1980, 1987), reflects primarily empirically observed clusters of symptoms and signs without identifying underlying disturbances in fundamental psychological processes. For example, in his presidential address at a meeting of the American Psychopathological Association, Bernard J. Carroll criticized DSM-III and proposed that the clinical symptoms and signs of bipolar affective disorder should be seen as manifestations of three basic psychopathological processes: one concerned with central pleasure-reward systems, one with central pain mechanisms, and one with psychomotor regulation mechanisms. As he noted, this was an elaboration and restatement of an earlier communication (Carroll 1983).

Quite apart from whether his specific proposal for bipolar affective disorder is correct, Carroll's general thesis is persuasive. We want to relate clinical phenomena to dysfunctions in basic brain systems that subserve psychological processes. Bodily physiology is manifested in a number of different systems, some reflecting organ function, others reflecting cellular functions across various organs. The brain manifests the same pattern, consisting as it does of parallel, distributed systems subsuming different functions, from basic vegetative activities to those dealing with perception, learning, and memory. Clinical disorders may reflect disturbances in different combinations of such systems, each with characteristic manifestations of disordered function. Thus, certain symptoms and signs may result from disturbances in system A, whereas others may result from disturbances in system B, and so on. It may be that Carroll's (1983) proposal, and that of others cited by him, is not yet part of the standard approach because the best way to categorize fundamental psychological functions remains unclear. As Carroll noted, his proposal is in response to the challenge from neuroscientists that psychologists and psychiatrists must strive to reduce the varied manifestations of psychiatric illness into basic elements that neurobiologists can try to correlate with dysfunction in specific brain systems. Such correlations would, of course, be powerful validators of the hypothesized basic psychological functions.

It is not surprising that the suggestions of Carroll and others have not yet been followed widely because we are only *now* beginning to see the possibility of recognizing and defining basic psychological function–brain system correlations. Without these correlations, many investigators have been hesitant (and would probably continue to be) about putting much weight on hypothesized fundamental psychological functions that cannot be anchored to specific neural networks. But the time may now be right. New and exciting work suggests that we may be on the threshold of discovering consistent confluences between specific brain networks and specific psychological functions, including pathological functions (or psychopathology). We can hope that these findings will soon open up new parameters for the classification of psychiatric disorders. If correlations between basic psychological functions and brain network systems can be clinically characterized and measured, they will become part of our diagnostic categories.

A similar evolution has taken place in general medical diagnoses. Progress generally begins with careful clinical evaluations of the patient's complaints and the findings on physical examination, coupled with the evaluation of responses to intervention or the natural course of the illness. As the symptoms and signs become identified with alterations in bodily physiology (through studies of urine, blood, and tissues), illnesses become better classified as disorders of the cardiovascular system, the respiratory system, the liver, or whatever locus is relevant. For example, with further research, it becomes possible to subdivide cardiovascular disorders into those associated with abnormalities in the coronary arteries or those related to disturbances in the structure of heart valves. Similarly, respiratory system disorders become divided into subtypes such as those with certain inflammatory and granulomatous lesions in the lung parenchyma, especially the upper lobes, and certain tumors of the larger bronchi.

Advances in general medicine were driven by discoveries in pathology and bacteriology as well as new technologies such as electrocardiography and X rays. Parallel advances in psychiatry may soon be feasible based on the introduction of powerful neuroimaging techniques and exciting discoveries in many areas of neurobiology.

Although most readers of this book are aware of and ready to accept these points, less than 15 years ago, the leaders of American psychiatry were not ready to accept the *central and essential role* of diagnosis in their thinking about the field and in their hope for progress through research. Many were actually opposed to a new emphasis on diagnosis. They justified their opposition to diagnosis in terms of the alleged risk to the clinician's recognition and acceptance of the uniqueness of each individual patient. Others were either indifferent or ambivalent, because they had not yet come to understand that the progress of scientific effort in psychiatry, as in the rest of medicine, is indissolubly linked to progress in diagnosis and classification.

As I have argued repeatedly, the uniqueness of each individual, sick or well, is not in dispute; it is a fact. But scientific progress depends on regularities and on identifying important common denominators. I believe that the argument about uniqueness was nearly always spurious. Resistance to the importance of diagnosis in psychiatry stemmed from seriously flawed conceptual thinking about scientific advances in the field of psychiatry. Such resistance reflected a failure to recognize that repeated, careful observations, delimited by measurable or countable criteria and presented to informed colleagues, are the fundamental scientific activities required to test concepts and hypotheses and to make comparisons of treatment results.

The view articulated here is that diagnosis (regarded as synonymous with classification) is an essential component of clinical research as well as clinical practice. Teaching and learning to do clinical research or to care for patients cannot proceed without diagnostic thinking. It was never possible. Those who seem to resist the deliberate emphasis on diagnosis are like the humorous fictional character who was surprised to discover that he was able to speak prose.

The questions are no longer "Do we need diagnostic systems?" or "How important is diagnosis?" Instead, the question is, "How can we improve diagnosis?" The correct answer is straightforward: well-designed, systematic, controlled studies of illness—whether epidemiological, clinical, physiological, biochemical, psychological, or pharmacological—will lead to improved diagnosis and diagnostic systems. Adherents of existing diagnostic categories and systems must strive to be open to criticism and suggested changes, especially when these suggestions are based on appropriate studies that explicitly or implicitly provide validation of the criticism and proposed changes.

If some colleagues wish to propose new parameters to be used for classification, they should be encouraged to carry out the appropriate studies. These should include efforts to define criteria for recognizing or measuring the new parameters as well as efforts to show their validity. Such scientific efforts will increase the likelihood that psychiatry as a discipline will advance more rapidly.

To restate my principal argument: diagnosis, including its conceptual foundation, is a sine qua non of scientific medicine and of scientific psychiatry.

References

American Psychiatric Association: Diagnostic and Statistical Manual of Mental Disorders, 3rd Edition. Washington, DC, American Psychiatric Association, 1980

American Psychiatric Association: Diagnostic and Statistical Manual of Mental Disorders, 3rd Edition, Revised. Washington, DC, American Psychiatric Association, 1987

Carroll BJ: Neurobiologic dimensions of depression and mania, in The Origins of Depression: Current Concepts and Approaches (Dahlem Konferenzen 1983). Edited by Angst J. Berlin, Springer-Verlag, 1983

Guze SB: Schizoaffective disorders, in Comprehensive Textbook of Psychiatry, 4th Edition. Edited by Kaplan HI, Sadock BJ. Baltimore, MD, Williams & Wilkins, 1984

2

The Relationship Between Anxiety and Depressive Disorders

Sir Martin Roth, M.D., F.R.C.P(sychiat), Sc.D.(Hon)
Christopher Mountjoy, M.B.B.S., F.R.C.P(sychiat)

That the emotions of anxiety and depression are related to each other is clear from observation of normal individuals, from introspection, from phenomenological investigations, and from studies undertaken in clinical psychiatric settings. Unitary views such as those of Mapother (1926) and Lewis (1936), which regarded states of anxiety as continuous with depressive disorders within the larger family of affective illnesses, continue to be influential. But evidence adduced in recent years—in relation to such conditions as "panic disorder" and agoraphobia, as well as other anxiety disorders and depressive states—has called the validity of these unitary views into question.

Goldberg and colleagues (Goldberg 1972; Goldberg and Huxley 1980; Goldberg et al. 1987), who have conducted extensive investigations into the prevalence and character of psychiatric disorders in the community, have concluded that anxiety-depression is the most common affective disorder encountered in primary care settings. Goldberg considers that the two components of the disorder constitute a single entity that is most validly conceived as a continuous dimension. These investigators used the 30-item General Health Questionnaire (GHQ-30) to examine a large community sample of 1,310 white subjects and 1,310 black subjects in the United States. The main factor was labeled a depression and anxiety factor and accounted for about 21% of the variance. Anxiety and depression as recorded with the GHQ-30 could not be separated from each other in any of the factor solutions attempted. However, in DSM-III (American Psychiatric Association 1980), as in all systems of classification that preceded it, anxiety states and depressive illnesses are treated as distinct categorical entities.

Charles Darwin was probably the first scientist to couple the emotions of depression and anxiety in a single account and to delineate both their separate roles and the relationship between them. In his "Expression of the Emotions in Man and Animals" (1872), he wrote

> After the mind suffers from a paroxysm of grief and the cause still continues, we fall into a state of low spirits or are utterly dejected. Prolonged pain generally leads to the same state of mind. If we expect to suffer we are anxious; if we have no hope of relief we despair.

It is plain that, when some ordeal that provokes anxiety is perceived as being beyond control, depression would likely supervene. It is equally evident that the depressed person who anticipates further despondency will probably suffer a complication of his or her affective state as a result of secondary anxiety. Such associations were implicit in Darwin's (1872) description, which merely implied some measure of correlation between the quite distinctive emotions to which he allocated distinct biological functions. In the case of emotional disorder observed in clinical or community settings, the extent to which anxiety disorders and depressive states can be differentiated has to be determined empirically.

In this chapter, we discuss separate investigations taken in sequence to resolve a succession of questions regarding the unity or diversity of the anxiety and depressive disorders. The first study suggested that differentiation was possible with the aid of multivariate statistical techniques. Later investigations were developed to determine the factors that had contributed to this separation and to the methods that could be devised to maximize it for the purposes of clinical diagnosis and scientific investigation.

These findings paved the way to a body of observations that demonstrated that separation of the depressive illnesses from anxiety disorders could be achieved with the use of personality measures and biographical and developmental data alone. The symptoms, signs, and behavior manifest in the presenting emotional disorder had been omitted from these analyses.

A Phenomenological Investigation and Follow-Up Study

The results of the first enquiry have been published in full (Gurney et al. 1972; Kerr et al. 1972, 1974; Roth et al. 1972; Schapira et al. 1972). Only brief reference will be made to them. One hundred forty-five patients were examined with the aid of a standardized interview and examination covering a wide range of biographical, developmental, and clinical items. All the features that characterize depressive and anxiety disorders were included in the examination schedule and were therefore assessed, graded, and recorded in each case. This procedure differs from that adopted in the unstandardized psychiatric examination. The course to be followed in an examination undertaken with a standardized schedule is preset to ensure that all features relevant for each of the diagnoses under inquiry are evaluated and graded.

Initial diagnoses allocated the patients either to an anxiety, depressive, or (for those in whom a definite diagnosis could not be made) intermediate group. Of the 145 patients studied, 68 were initially diagnosed as suffering from an anxiety or phobic state, 62 were diagnosed as depressive, and 15 were allocated to a "doubtful" group. The distribution of patients' summated component scores proved to be unimodal. A discriminant function analysis (Gurney et al. 1972) achieved clear separation between the two groups, with the two modes corresponding to the anxiety states and depressive disorders. A principal components analysis was carried out on 58 items, which included some personality features and data relating to history and premorbid adjustment. The first component contrasted depressive features at one pole with anxiety features at the opposite pole. A discriminant function analysis of patients' scores on the first component achieved clear separation between the anxiety and depressive groups. Some clinical features were, however, common to both groups. A number of anxiety features were present in depressive disorders, whereas depression of variable character and suicidal tendencies were present in a proportion of anxiety and phobic states, although there were clear differences between the groups in prevalence of such features.

Such a finding provides merely tentative evidence in favor of the dualistic view, particularly because the results obtained with the aid of multiple regression techniques require replication in further samples before they can be accepted in confirmation of the initial main hypothesis. (The results of further studies will be described at a later stage.) However, independent evidence was adduced, from follow-up investigations conducted by researchers unaware of the original diagnosis. Confirmation for the dualistic concept of the disorders was provided by several lines of evidence.

- A follow-up study of course and outcome of the original hypothetical groups by investigators ignorant of the initial diagnoses demonstrated that at each stage of the investigation over a 3- to 4-year period, a significantly higher proportion of depressed patients had improved or recovered than the anxious and phobic patients (Kerr et al. 1972, 1974).
- Very little crossover occurred between the two main groups in the longitudinal studies. Anxious patients tended, on relapse, to manifest anxiety neuroses, and depressive patients to develop a depressive illness. It should be added that, in most cases, agoraphobic patients and those with social or other phobias exhibited recrudescence of their specific symptoms when they relapsed.
- The results achieved with the aid of the physical treatments, prescribed by a number of consultants unconnected with the inquiry, showed that

depressed patients responded significantly better than anxious patients to both electroconvulsive therapy (ECT) and tricyclic antidepressants (Gurney et al. 1970).

- The clinical features present during illness and the antecedent life events and personality features that predicted outcome in anxious patients were markedly different from those that predicted outcome in depressive patients (Kerr et al. 1974).
- The other studies summarized in this chapter were submitted to further test the hypothesis that the depressive disorders and anxiety states are distinct from each other, with little overlap, and have attempted to sharpen the clinical criteria and procedures that may reliably differentiate them from each other.

An Investigation Confined to the Features of the Presenting Clinical Picture

The first of these studies concerned 117 inpatients and outpatients with a wide range of severity (46 men and 71 women) (Mountjoy and Roth 1982a, 1982b). Forty-three subjects were given a tentative diagnosis of depressive illness, 44 of anxiety neurosis, and 30 of phobic neurosis. To exclude the possibility that the separation of syndromes previously reported had been largely due to the inclusion of endogenous bipolar and unipolar cases, with their sharply defined features, patients with endogenous depression were not included. Only subjects with nonendogenous depressions and anxiety and phobic states were submitted to investigation.

Administration of Rating Scales for Depression and Anxiety

Two methods of clinical evaluation and measurement were used. The first consisted of the administration of a range of widely used rating scales for depression and anxiety: the Hamilton Anxiety Rating Scale (Hamilton 1959), the Hamilton Depression Scale (Hamilton 1960), the Agoraphobic Rating Scale (Gelder and Marks 1966), a global measure of severity, the Wakefield Modification of the Zung Self-Rating Scale for Depression (Snaith et al. 1971), the Anxiety Self Rating Scale (Lipsedge et al. 1971), and the Newcastle Anxiety-Depression Rating Scale (Gurney et al. 1972).

Because each test confines clinical inquiries to specially formulated questions that have to be scored in specific ways, these tests provided a means of investigating the problem on which the inquiry was focused, with instruments relatively free from bias.

A principal components analysis of the total scores on each of the rating scales yielded two main factors. The first was a general severity factor, which accounted for 43.7% of the variance, and the second a bipolar factor, which accounted for 21.3% of the variance. This factor separated the rating scales into two groups and was clearly an anxiety-depression dimension. When patients' summated scores on the different tests were plotted on this distribution, the component proved to be an anxiety-depression factor separating the two patient groups, with the depressive cases at one end and the anxiety cases at the other. A measure of overlap occurred between the distribution of the two groups.

Analysis of Individual Rating Scales

A stepwise discriminant function analysis of the second component included in the first stage only those patients in whom clinical diagnosis and allocation by the Newcastle scale had yielded identical results. With the aid of the discriminant function coefficients elicited for all the scales, the entire patient population was plotted, including "doubtful" cases. The definite cases were those in whom diagnoses and allocation on the basis of the Newcastle scale had yielded identical results, and the "doubtful" cases were those in whom the results had been discrepant. With the aid of coefficients derived from the discriminant function, 97.7% of cases were correctly assigned to the groups by which they had been assigned by a definite diagnosis at the outset. The most powerful discriminators were the Hamilton Depression Scale (Hamilton 1960) and the Depression Self-Rating Scale (Snaith et al. 1971) in the depressive half of the distribution and the Newcastle Anxiety-Depression Rating Scale and the scale in the other half of the distribution. We must stress, however, that separation achieved by this analysis was not identical with the clinical diagnosis initially made. Thus, when account was also taken of the results achieved in the "doubtful" class, the overall misclassification was 16.24%.

One of our most interesting findings was the extent to which the differentiation could be achieved with the aid of a discriminant function analysis of the items of any one of the individual rating scales, such as the Hamilton Depression Scale (Hamilton 1960), which was designed purely as a measure of severity. Not only the Newcastle Anxiety-Depression Rating Scale but each of the other tests proved, to a greater or lesser extent, to discriminate between anxiety and depressive states.

Using the Hamilton Depression Scale alone, the entire group of cases was correctly classified in 89.9% of cases. Analysis of the scores on this test resulted in a misclassification rate of 11%, an unexpected result for a scale intended to measure the severity of depressive affect alone. The statistical weights for a number of features carried opposite signs. Depression of mood and insomnia in one direction and tension and irrational fears in the other were the most powerful discriminators. The Self-Rating Scales were the least successful in discrimination. Overall, the items from the scales were those delineating depression, suicide, insomnia, and weeping on the one hand and feelings of tension, panic, fears and somatic anxiety on the other.

Taken together, these results demonstrate that a number of reliable and validated scales can be used to assist clinical discrimination in that large territory in which clear landmarks do not yet exist for affective disorders.

Analysis of Clinical Items

The second technique involved analyses of separate clinical items. A principal components analysis was carried out on the clinical items recorded on a standard form for 108 of the same patients administered the rating scales. The analysis yielded two clinically meaningful components. The first accounted for 12% of the variance and was in part a general component. The second factor, accounting for 9% of the variance, was clearly bipolar. For the first component the most important discriminators, with negative (depressive) weighting, were depressed mood and pessimistic outlook, whereas reactivity of depression and increased physiological responses carried positive (anxiety) weighting. In the second component, depressed mood and pessimistic outlook were the highest negative (depressive) discriminators; the highest positive (anxiety) ones were increased physiological responses, situational phobias, panic attacks, derealization, fear of being left alone, and compulsive phenomena.

Using the 21 items with the highest loadings on the principal components analysis, two discriminant function analyses were performed with the aid of the Mahalonobis technique. Reference will be made here only to the second of these. The patients who did not fall into the appropriate side of the distribution of the summated scores on the principal components were judged to be the "doubtful" cases. The analysis was carried out in the total group and also in two randomly derived subgroups. The population of 108 patients was clearly partitioned into two distinct groups (Figure 2–1). Once again, when the distribution of patients according to clinical diagnosis was superimposed on the distribution of the discriminant function scores, classification of patient groups did not yield results identical with the subdivision made initially on the strength of clinical criteria. The misclassification rate for component one proved to be 17.59% and for component two, 18.25% (Mountjoy and Roth 1982b).

In summary, the findings confirmed (with the aid of different techniques, including a number of objective rating scales) the findings of Prusoff and Klerman (1974), Downing and Rickels (1974), Rickels and colleagues (1979), and Huppert and colleagues (1989), all of which adduced evidence favoring a relatively clear distinction between depressive and anxiety disorders.

Measures of Personality and the Classification of Affective Disorder

Information concerning premorbid personality and the developmental history of the patients we studied was recorded systematically at the time of initial interview. The items used were those recorded for the original Newcastle sample and the definitions were those reported by Roth and colleagues (1972). Because the number of variables was greater than the number of cases, the number of items was reduced from 65 to 37 by eliminating any item that occurred with a frequency of more that 90% or less than 10%. The remaining variables were further reduced by doing three separate discriminant function analyses and selecting the best discriminants from each analysis.

In discriminant function analysis using the Mahalonobis technique, definite cases were defined as those in which clinical diagnosis and the diagnosis formed from the Newcastle Anxiety-Depressive Diagnostic Scale were congruent; incongruent diagnoses were regarded as "doubtful." Statistically significant separation ($P = 0.0004$) was achieved between the depressive group and the anxious and phobic group, with a 78% correct classification of the grouped cases. The greatest negative loadings were for impulsive behavior and childhood temper tantrums, and the greatest positive loadings were the Maudsley Personality Inventory (MPI) neuroticism score (Eysenck 1959) and premorbid somatic anxiety. Separate analysis for men and women produced different results. A statistically significant separation was not achieved for men ($P = 0.085$) but was highly significant for women ($P = 0.0005$); 87% of grouped cases were correctly classified. The same negative items were found as for the total group, but the highest positive loadings for women were evanescent enthusiasms, childhood bed-wetting, and nail biting.

Only brief reference can be made to an investigation (Caetano 1981; Roth et al. 1982) into 152 patients covering the entire range of disorders of affect, including endogenous and neurotic depressive patients, as well as patients with general anxiety states, agoraphobia, and other phobic anxiety disorders. Patients were investigated with the aid of the Present State Examination (Wing et al. 1974); data were analyzed using multivariate techniques. Although the results were largely in accord with those of the investigations already described, the study's main interest in the present context is in the findings that issued from personality tests.

Personality ratings of each patient were derived with the aid of the MPI and the Cattell 16 PF battery (Cattell et al. 1970). There was 84% agreement between the classification of patients into groups of definite "anxiety state" and "depressive disorder" and the separation of the 152 patients achieved with the aid of the personality ratings. Those patients with anxiety states emerged from this part of the investigation as "shy," "introverted," and "anx-

Figure 2–1. Distribution of anxious and depressed groups on discriminant function of rating scales. Hatched area indicates the subjects allocated to the depressive discriminant function. (From Roth and Mountjoy 1982; reprinted with permission.)

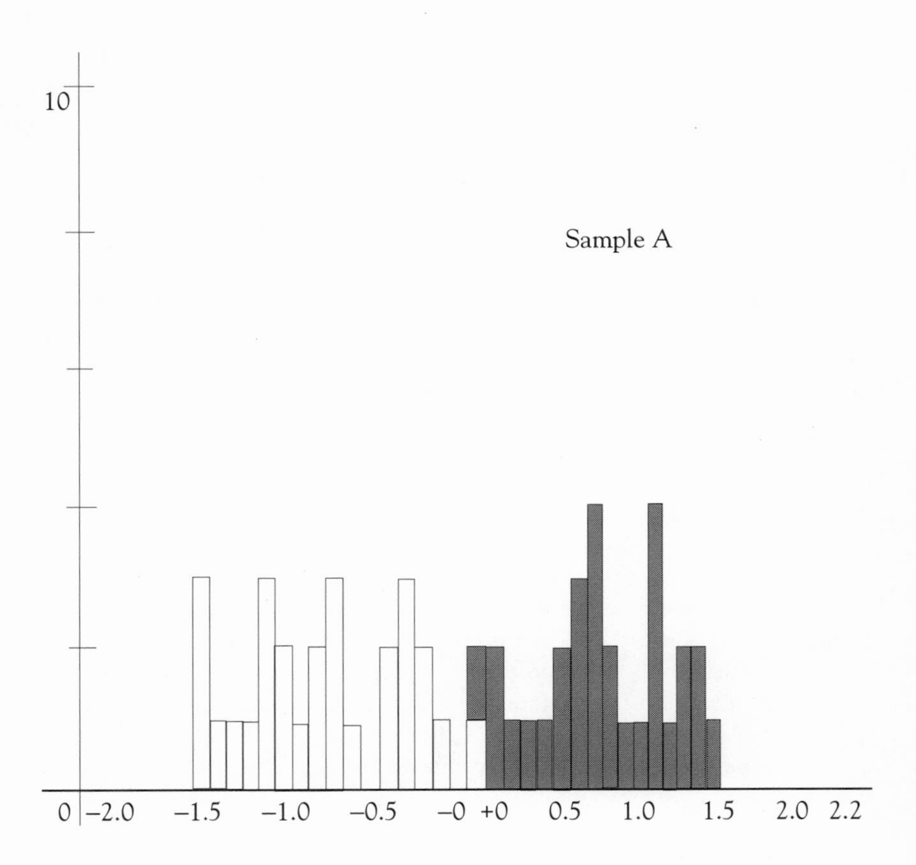

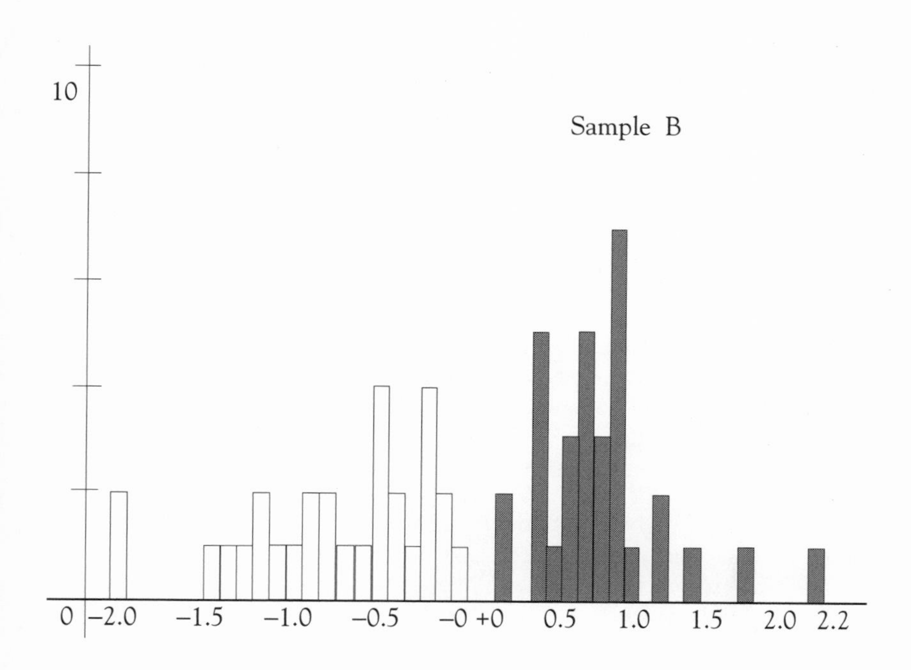
10
Sample B
0
–2.0
–1.5
–1.0
–0.5
–0 +0
0.5
1.0
1.5
2.0
2.2

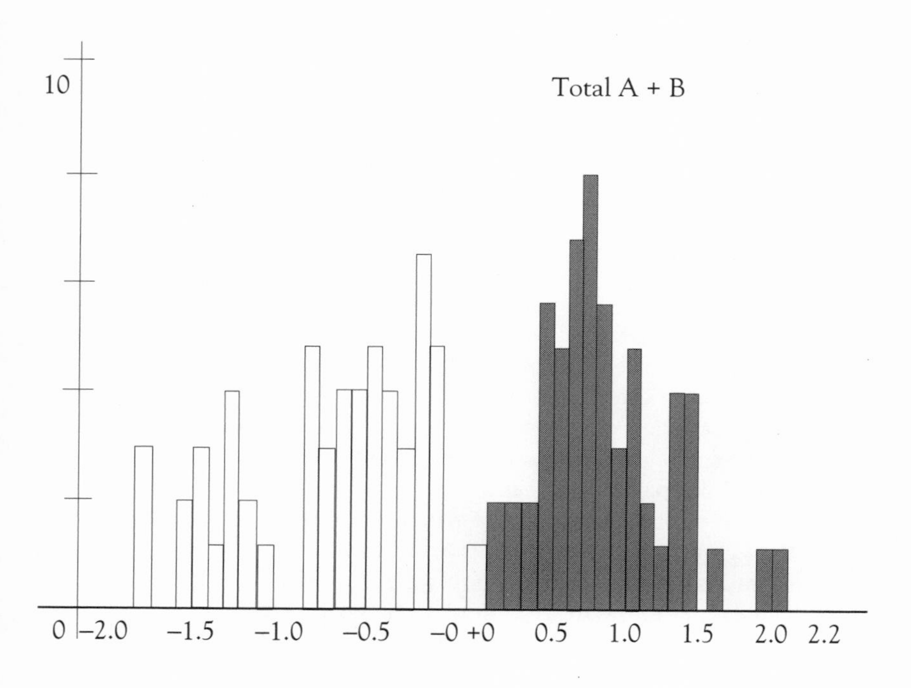
10
Total A + B
0
–2.0
–1.5
–1.0
–0.5
–0 +0
0.5
1.0
1.5
2.0
2.2

iety prone," whereas those from the depressive group proved to be "outgoing," "emotionally stable," and "tender minded." As in previous investigations (Gurney et al. 1972), those patients with depressive types of illness obtained lower scores on neuroticism as defined by the MPI than those with anxiety states.

Finally, with the aid of canonical variate analysis on the MPI and 16 PF scores, the two canonical variates derived succeeded in separating 73% of the patients into three groups that corresponded to the following syndromes:

- A group of overlapping syndromes: "psychotic depression," "retarded-agitated depression," and "endogenous depression," which had been derived from computer analysis using the "Catego" program.
- "Neurotic depression."
- "Anxiety state," including both phobic and simple anxiety disorders.

Personality data obtained independently were able to classify 84.3% of the patients correctly into the two groups, anxiety neurosis and depressive disorder, into which they had been initially allocated. Needless to say, personality tests are incapable of achieving this degree of diagnostic accuracy in unselected groups of psychiatric patients. But if a practitioner is presented with a group of patients previously diagnosed clinically as suffering from predominantly anxious, depressive, or both kinds of symptoms without any further specification, the use of these tests appears to provide a considerable measure of independent discriminating power.

Discussion

Correctly solving the riddle posed by the relationship between anxiety and depressive disorders is important for a number of reasons. When two relatively distinct clinical syndromes are erroneously treated as a unit, neurobiological features that are specific for one of the syndromes may elude detection through dilution of the patient population selected for inquiry by patients who suffer from an unrelated disorder. Inquiries into the efficacy of new treatments may suffer for similar reasons.

The investigation of heterogeneous samples in controlled clinical trials may yield insignificant differences between the active compound and a placebo or some alternative treatment.

Although allocation to the two patient groups will have to be random where the samples containing a mixture of two types of patient are treated as a single group, the patients receiving active substance or placebo, respectively, could differ in the relative proportion of patients drawn from the various subgroups. One could generalize more widely from these two examples to other situations in which significant findings may be obscured through operation of the dilution factor.

The accurate choice of treatment in clinical practice also depends on the validity of the classification employed. This view might be called into question in the light of results recorded in trials of pharmacological treatment in depressive and anxiety disorders. But a close examination of the results reported proves such a conclusion to be unjustifiable. In investigations of patients with depressive and anxiety symptoms who were treated with amitriptyline, diazepam, or a combination of the two substances in one case (Johnstone et al. 1980) and phenelzine or amitriptyline in the other (Paykel et al. 1982), each drug or combination of drugs proved effective in the amelioration of anxiety and depression. But there was no significantly different influence on either affect.

The fact that isolated anxiety and depressive symptoms were not differently influenced is hardly surprising. These results leave unanswered the question as to whether or not some pharmacological or other forms of treatment are significantly more or less beneficial in depressive *syndromes* than in anxiety *syndromes*.

There is a body of evidence that certain drugs that are indubitably effective in depression are also efficacious in panic disorder and agoraphobia (Klein and Rabkin 1981; Zitrin et al. 1981). However, the view that there is something specific about the action of tricyclic compounds in agoraphobic and related disorders is called into question not only by the efficacy of the triazolobenzodiazapine alprazolam, but also by the reported effectiveness of high doses of diazepam in panic disorder (Noyes et al. 1984). There is also an overlap in the therapeutic efficacy of monoamine oxidase (MAO) inhibitors between agoraphobia and depressive disorders. Aside the fact that all these drugs ameliorate rather than eradicate agoraphobia, which proves to be a chronic disorder in a high proportion of cases, the similarities of treatment response discussed here do not justify confident inferences regarding the unity or separateness of the anxiety or depressive disorders. The range of disorders influenced by corticoids provides a poor foundation for judgments about classification of the disorders in question.

There is, however, one form of treatment that discriminates sharply between depressive and anxiety disorders. ECT is the only form of treatment proven effective in psychotic and some endogenous depressions and also in a substantial minority of severe and suicidal neurotic depressions. However, it is ineffective in anxiety disorders of every kind and often aggravates the symptoms in patients with these disorders.

In the majority of factor analytic and related forms of investigation of the relationship between anxiety disorder and the depressive illnesses, the two have been found to be distinct (Mullaney 1984). These findings have been confirmed by the investigations reviewed here. The evidence offered in the analysis of results from a range of rating scales for anxiety and depression has provided fresh testimony for the dichotomous view. The range of severity of the disorders studied was wide, and the subjects included many outpatients. A community sample comprising patients from general practice (Fahy et al. 1969) yielded closely similar results.

Another feature of the inquiries summarized here has been the demonstration that the data derived from history of the development of adjustment and personality features can be of considerable value for discrimination in diagnosis. This may appear at variance with the results of the first stage investigation (Mountjoy and Roth 1982a, 1982b), which partitioned the two conditions ostensibly on the basis of cross-sectional examination of the present mental state alone. This impression may be illusory to a certain extent, in that the initial examination devoted a considerable amount of time to studying the history of the premorbid adjustment in various roles and to systematic evaluation of personality, although these were not included in the statistical analyses. However, this is unlikely to explain or to invalidate the clear partition achieved with the aid of personality and developmental data independently of the findings derived from examination of the patient's presenting mental state.

The view that the antecedent personality profile and the developmental histories of patients with anxiety and other neurotic states are an integral part of the clinical profile of the neuroses and are closely related to differences in their specific manifestations was a basic principle in the thought and teaching of the pioneers of descriptive psychiatry. For Kurt Schneider (1950), neuroses were lifelong characteristics of the personality, and no clear distinction was made between neurotic illness and disordered personality ("psychopathy"). The influential Swedish psychiatrist Sjobring (1973) took an essentially similar view. Slater's concept of the "neurotic constitution" (Slater 1943) drew upon observations that showed continuity between the clinical features of the different neuroses and the features of the premorbid personality in which they occurred. He also adduced evidence for a close causal relationship between them. Contemporary workers such as Gray (1983) and Cloninger (1986) allocate a role of basic importance to personality dimensions in the development of anxiety and other neurotic disorders.

DSM-III provides a separate dimension (Axis III) for the recording of personality disorders and personality traits. However, in clinical research publications, it is rare for other than Axis I diagnoses to be presented in definitions of the patients under investigation. Axis II diagnoses rarely receive mention in biological investigations in particular, possibly because they are of rather poor reliability. The implicit assumption is that the axes are orthogonal and hence independent. The failure to use personality and related data for characterizing patients in the affective and anxiety related disorders may be responsible for the marked areas of overlap recorded between them in some studies and the many examples of "comorbidity" or coexistence of two or more syndromes that have been uncovered in the course of studies in this area with the aid of DSM-III-R (American Psychiatric Association 1987.

The evidence that anxiety and depressive disorders are relatively distinct does not signify that the effects of anxiety and depression and the illnesses in which they predominate are wholly unrelated. This is clearly not the case; there are lines of continuity as well as lines of cleavage between them. Those afflicted with anxiety for long periods are liable to become depressed, and severely depressed patients often have symptoms of anxiety. This does not signify that anxiety and depressive disorders constitute one single syndrome, or that double diagnoses are in order when they share some of each other's features.

Both the continuities and the lines of separation between anxiety and depressive disorders must be borne in mind in clinical practice and investigated by scientific means. Such studies should eventually shed light on the biological origins of these affective states and pave the way for our deeper knowledge of their underlying pathophysiology and causation.

References

American Psychiatric Association: Diagnostic and Statistical Manual of Mental Disorders, 3rd Edition. Washington, DC, American Psychiatric Association, 1980

American Psychiatric Association: Diagnostic and Statistical Manual of Mental Disorders, 3rd Edition, Revised. Washington, DC, American Psychiatric Association, 1987

Caetano D: The classification of affective disorders: the relationship between anxiety states and depressive disorders. Doctoral dissertation, University of Cambridge, Cambridge, England, 1981

Cattell RB, Eber HW, Tatsuoka MM: The Handbook for the Sixteen Personality Factor Questionnaire. Champaign, IL, Institute for Personality and Ability Testing, 1970

Cloninger CR: A unified biosocial theory of personality and its role in the development of anxiety states. Psychiatr Dev 3:167–226, 1986

Darwin C: The Expression of the Emotions in Man and Animals. London, Journal of Murray, 1872

Downing RW, Rickels R: Mixed anxiety-depression, fact or myth? Arch Gen Psychiatry 30:312–317, 1974

Eysenck HM: Manual of the Maudsley Personality Inventory. London, University of London Press, 1959

Fahy TJ, Brandon S, Garside RF: Clinical syndromes in a sample of depressed patients: a general practice material. Proceedings of the Royal Society of Medicine 62:331–335, 1969

Gelder MG, Marks IM: Severe agoraphobia—a controlled prospective trial of behavior therapy. Br J Psychiatry 112:309–319, 1966

Goldberg DP: The Detection of Psychiatric Illness by Questionnaire. London, Oxford University Press, 1972

Goldberg DP, Huxley P: Mental Illness in the Community. The Pathway to Psychiatric Care. London, Tavistock Publications Ltd, 1980

Goldberg DP, Bridges K, Duncan-Jones P, et al: Dimensions of neuroses seen in primary care. Psychol Med 17:461–470, 1987

Gray JA: Anxiety, personality and the brain, in Physiological Correlates of Human Behavior, Vol III. Edited by Gale H, Edwards JA. New York, Academic, 1983

Gurney C, Hall R, Harper M, et al: The Newcastle Index. Lancet 2:1275–1378, 1970

Gurney C, Roth M, Garside RF, et al: Studies in the classification of affective disorders. The relationship between anxiety and depressive illness, II. Br J Psychiatry 121:162–166, 1972

Hamilton M: The assessment of anxiety states by rating. Br J Med Psychol 32:50–55, 1959

Hamilton M: A rating scale for depression. J Neurol Neurosurg Psychiatry 23:56–61, 1960

Huppert FA, Walters DE, Day NE, et al: The factor structure of the General Health Questionnaire (GHQ-30): a reliability study on 6,317 community residents. Br J Psychiatry 155:178–185, 1989

Johnstone EC, Cunningham Owens DG, Frith CD, et al: Neurotic illness and its response to anxiolytic and antidepressant treatment. Psychol Med 10:321, 1980

Kerr TA, Roth M, Schapira K, et al: The assessment and prediction of outcome in affective disorders. Br J Psychiatry 121:167–174, 1972

Kerr TA, Roth M, Schapira K: Prediction of outcome in anxiety states and depressive illnesses. Br J Psychiatry 124:125–133, 1974

Klein DF, Rabkin JG: Anxiety: New Research and Changing Concepts. New York, Raven, 1981

Lewis AJ: Melancholia: prognostic studies and case material. Journal of Mental Science 82:488–558, 1936

Lipsedge MS, Rees WL, Pike DJ: A double blind comparison of dothiepin and amitriptyline for the treatment of depression with anxiety. Psychopharmacologica 19:153–162, 1971

Mapother E: Discussion on manic-depressive psychosis. BMJ 2:872–876, 1926

Mountjoy CQ, Roth M: Studies in the relationship between depressive disorders and anxiety states, I: rating scales. J Affective Disord 4:127–147, 1982a

Mountjoy CQ, Roth M: Studies in the relationship of depressive disorders and anxiety states, II: clinical items. J Affective Disord 4:149–164, 1982b

Mullaney JA: The relationship between anxiety and depression: a review of some principal component analytic studies. J Affective Disord 7:139–148, 1984

Noyes R Jr, Anderson DJ, Clancey J, et al: Diazepam and propranolol in panic disorder and agoraphobia. Arch Gen Psychiatry 41:287–292, 1984

Paykel ES, Rowan PR, Rao BM, et al: Atypical depression: nosology and response to antidepressants, in Treatment of Depression: Old Controversies and New Approaches. Edited by Clayton P, Barrett J. New York, Raven, 1982

Prusoff G, Klerman GL: Differentiating depressed from anxious neurotic outpatients. Arch Gen Psychiatry 30:302–308, 1974

Rickels K, Downing RW, Stein M: The differentiation between neurotic depression and anxiety and anxiety and drug treatment, in Neuropsychopharmacology: Proceedings of the 11th Congress of the Collegium Internationale Neuropsychopharmacologicum, Vienna 1978. Edited by Saletu B, Berner P, Hollister L. Oxford, Pergamon, 1979

Roth M, Mountjoy CQ: Handbook of Affective Disorders, Chapter 6. Edited by Paykel ES. New York, Churchill Livingstone, 1982

Roth M, Gurney C, Garside RF, et al: Studies in the classification of affective disorders: the relationship between anxiety states and depressive illness, I. Br J Psychiatry 121:147–162, 1972

Roth M, Mountjoy CQ, Caetano D: Further investigations into the relationship between depressive disorders and anxiety states. Pharmacopsychiatry 15:135–141, 1982

Schapira K, Roth M, Kerr TA, et al: The prognosis of affective disorders—The differentiation of anxiety states from depressive illness. Br J Psychiatry 121:175–181, 1972

Schneider K: Letter from Germany: systematic psychiatry. Am J Psychiatry 107:334, 1950

Sjobring H: Personality structure and development: a model and its application. Acta Psychiatr Scand Suppl 244, 1973

Slater E: The neurotic constitution: a statistical study of 2,000 neurotic soldiers. Journal of Neurology and Psychiatry 6:1–16, 1943

Snaith RP, Ahmed SN, Mehta S, et al: Assessment of the severity of primary depressive illness: Wakefield self-assessment depression inventory. Psychol Med 1:143–149, 1971

Wing JK, Cooper JE, Sartorius N: The Description and Classification of Psychiatric Symptoms; An Instruction Manual for the P.S.E. and Catego Systems. London, Cambridge University Press, 1974

Zitrin CM, Woerner MG, Klein DF: Differentiation of panic anxiety from anticipatory anxiety and avoidance behavior, in Anxiety, New Research and Changing Concepts. Edited by Klein DF, Rabkin JG. New York, Raven, 1981

3

The Schizoaffective Continuum

George Winokur, M.D.
Donald W. Black, M.D.
Amelia Nasrallah, M.S.

Because there are 24 or 25 different definitions of schizoaffective disorder, the evolutionary line in the development of the schizoaffective diagnosis is not easily traced. Several definitions are operational and well described (Levinston and Levitt 1987; Marneros and Tsuang 1986), whereas others encompass illnesses that share features with operational definitions of schizoaffective disorder, such as benign stupors, reactive psychoses, good premorbid schizophrenia, cycloid psychoses, schizophreniform illnesses, remitting schizophrenia, and acute schizophrenia (Winokur 1984). Certainly, 24 or 25 definitions represent a rough estimate, and we made no systematic effort to obtain all of the possible definitions and related diagnoses. Assuming 24 possible definitions and criteria sets, it would be possible to study any patient group and come up with 48 different studies (provided that schizoaffective mania and schizoaffective depression are considered separate disorders).

The number of possible studies in which an investigator might compare one definition with another is, of course, astronomical. Assuming all 48 definitions are independent, 1,128 comparative studies are possible. There would be 276 possible comparisons just for the schizoaffective depressive group and

---

This is the second of a triad of papers on the appropriate classification of schizoaffective disorder. The first paper is Winokur G, Nasrallah A: The schizoaffective continuum: nonpsychotic, mood congruent, mood incongruent, in Affective and Schizoaffective Disorders. Edited by Marneros A, Tsuang M. Berlin, SpringerVerlag, 1990. The third paper is Winokur G: The schizoaffective continuum. Annals of Clinical Psychiatry 1:19–24, 1989.

an equal number for the schizoaffective manic group. Is it any wonder that we have considerable problems with the concept of schizoaffective disorder? This fact is simply a monument to the fecundity of psychiatry's ability to come up with new definitions.

In the United States, essential elements of schizoaffective disorder are captured in the *Research Diagnostic Criteria* (RDC; Spitzer et al. 1978). Patients with this diagnosis have an episode of mania or depression with a specified number of affective symptoms concomitantly with specific psychotic features, such as mood-incongruent delusions or hallucinations. Patients satisfy the criteria if they have experienced a time during which affective symptoms are absent while schizophrenic symptoms such as those noted above are present; this criterion is harder to document. The diagnosis of schizoaffective disorder in the American standard nomenclature (DSM-III-R; American Psychiatric Association 1987) contains no rigorous criteria such as these of the RDC. The diagnoses of the affective disorders are broader in DSM-III-R than they are in the RDC; consequently, some patients who received the diagnosis of schizoaffective disorder based on the RDC would receive the diagnosis of affective disorder using DSM-III-R.

Several years ago, our research group used the RDC to evaluate a group of schizoaffective manic patients. We compared the schizoaffective manic patients with manic patients who had no schizophrenic-type symptoms and found no difference in course of illness or family history. These findings suggested that the schizoaffective manic patients are simply manic patients with a somewhat different clinical background. An additional finding was the possible association of a schizoaffective clinical picture in manic patients with endocrine abnormalities or a postpartum state (Winokur et al. 1986).

In a new and large data set, we have chosen to evaluate patients on a continuum. Nonpsychotic patients with depression and mania are being compared to manic or depressive psychotic patients with mood-congruent psychotic symptoms and to manic or depressive patients with mood-incongruent psychotic symptoms. These three groups of depressive or manic patients are being compared to determine how they may be similar in clinical picture, course of illness, and response to treatment.

Methodology

Two thousand fifty-four patients with an affective disorder were admitted to the University of Iowa Psychiatric Hospital from January 1, 1970, to December 31, 1981. These patients had chart diagnoses of unipolar depression, manic disorder, involutional melancholia, manic-depressive psychosis, atypical depression, atypical manic disorder, atypical psychosis, schizoaffective disorder, secondary depression, neurotic depression, cyclothymia, or dysthymia.

All archival material was systematically evaluated according to certain patient criteria, including sex, marital status, age at index admission, previous hospitalization, age at index hospitalization, duration of illness at admission, precipitating events, organic features at admission, suicidal thoughts at admission, melancholia symptoms at admission, delusions and hallucinations, treatment and response for index hospitalization, previous treatments and response, previous suicide attempts, outcome at discharge, and death (including cause of death). The index admission was the last admission if there was more than one. Relapse after discharge was recorded; the patient was considered to have relapsed if, on a subsequent clinic visit, he or she reported three or more depressive symptoms from a symptom checklist or if a psychiatrist diagnosed a manic syndrome. Suicide attempts were recorded if either serious or nonserious suicide attempts had been cited in the progress notes.

We also recorded the presence of a nonaffective psychiatric illness that might have predated the depression for which a patient was diagnosed and admitted (secondary depression). In this study, depression was considered secondary if a severe or life-threatening medical illness or a nonaffective psychiatric condition that predated or paralleled the depression was noted in the chart. No secondary depressive patients were included in this study.

Our methodology has been fully described previously (Black et al. 1987a, 1987b; Winokur et al. 1988). Essentially, we looked at patients with a DSM-III diagnosis (American Psychiatric Association 1980) of major depressive disorder or mania. Our sample included 604 unipolar depressive patients without psychotic symptoms, 76 unipolar depressive patients with mood-congruent psychotic symptoms, and 60 unipolar depressive patients with mood-incongruent psychotic symptoms. (We used the terms "mood-congruent" and "mood-incongruent" as defined in DSM-III-R.) The other large category included 188 nonpsychotic manic patients, 113 mood-congruent psychotic manic patients, and 88 mood-incongruent psychotic manic patients. The depressive patients and the manic patients were evaluated separately using a $2 \times K\ \chi^2$ analysis (df = 2).

Our method looked at severity in a continuous and systematic way. The unipolar depressive patients and the manic patients with mood-incongruent symptoms were considered to be the patients who most clearly represented schizoaffective disorder according to the RDC. They were then compared to two groups with separate levels of severity: those with unipolar depression and congruent psychotic symptoms were presumed to be more severely ill than those without psychotic symptoms. We considered this true for the manic patients as well. Psychotic symptoms in this study included delusions and hallucinations.

Finally, the method lent itself to a new possibility for analysis. We compared the unipolar depressive patients with incongruent psychotic symptoms to the manic patients with incongruent psychotic symptoms to see whether or not they paralleled previously reported differences in patterns between nonpsychotic unipolar patients and manic patients.

Results

Comparisons among unipolar depressive patients are shown in Table 3–1. Acute onset (duration of illness at admission) does not differentiate the three groups, but the mood-incongruent psychotic patients were less likely to have a chronic course prior to hospitalization (Table 3–1). Only 22% of the mood-incongruent group had been ill for 6 months before entering the hospital as opposed to 32% of the mood-congruent group and 40% of the group with no psychotic symptoms. A simple explanation for this difference might be that the mood-incongruent psychotic patients have aberrant behavior that is not tolerated by their families; therefore, they are brought to medical attention earlier.

Precipitating events occurred more frequently in unipolar depressive patients without psychotic symptoms. It is entirely likely that this large group of patients contained many individuals who meet the diagnosis of "neurotic depression," which, in general, is associated with a lack of psychotic symptoms and a positive report for precipitating factors by the patient (Winokur 1985; Winokur et al. 1987).

Table 3–1 shows that hallucinations were more frequent in the unipolar depressive patients with mood-incongruent psychotic symptoms. Because both unipolar depressive psychotic groups have delusions, this is a striking finding. If one accepts the possibility that hallucinations are a sign of increased severity, our data support the theory that unipolar depressive patients with mood-incongruent psychotic symptoms are the more severely ill group. Response to treatment is no different among the three groups. All patients were equally likely to respond to electroconvulsive therapy (ECT) and antidepressant medication. The mood-incongruent psychotic group was more likely to require antipsychotic drug maintenance, which is expected. Duration of follow-up was the same for all groups, and we found no significant difference occurred in the number of patients in each group who relapsed. Although the data suggested that the unipolar depressive patients with congruent psychotic symptoms were more likely to relapse, this difference was not statistically significant. Nor was there a significant difference in number of suicides at follow-up. Deaths from natural causes were less frequent in the mood-incongruent group (5%) than in the other groups (10% and 14% for nonpsychotic patients and mood-congruent psychotic patients, respectively). However, the mood-incongruent group was younger at index, which probably accounts for the difference.

Table 3–1. A comparison of nonpsychotic unipolar depressive patients versus depressive patients with congruent or incongruent psychotic symptoms

	Unipolar depressive patients			
Criterion	Without psychotic symptoms	With congruent psychotic symptoms	With incongruent psychotic symptoms	*P*
N	604	76	60	
Median age at admission	44	56	34	.02
% female	64	68	73	NS
< 4 weeks	50%	38%	35%	.001
7 or more hospitalizations	4%	4%	14%	.004
Mean age first ill	37	32	31	.02
Duration of illness at admission ≤ 4 weeks	9%	9%	12%	NS
Previous suicide attempts	33%	26%	35%	NS
Organic factors	14	28	19	.02
Hallucinations	0	17	50	.001
ECT given	44	67	50	.01
Marked improvement				
At discharge	61	68	57	NS
With ECT	59	56	70	NS
(or improvement) on antidepressants	19	14	19	NS
DST nonsuppression at 4 P.M. (≥5 mg/dl)	42	81	47	.01
Antipsychotic maintenance	21	35	57	.001
Diagnosis positive				
Melancholia	12	41	13	.001
Schizoaffective illness	0	0	45	
Follow-up, 2+ years	42	45	41	NS
No relapse	50	31	52	NS
Suicide attempt	5	4	3	NS
Suicide	4	5	5	NS
Death from other cause	10	14	5	NS

Table 3–2 presents the comparisons between nonpsychotic manic, mood-congruent manic, and mood-incongruent manic patients. The data in Table 3–2 are quite striking for their absence of significant differences among patients for criteria such as age, sex, marital status, length of hospitalization,

Table 3–2. A comparison of nonpsychotic manic patients and manic patients with mood-congruent or mood-incongruent psychotic symptoms

Criterion	Nonpsychotic manic patients	Psychotic manic patients: Mood-congruent	Psychotic manic patients: Mood-incongruent	*P*
N	188	133	88	
Age, median (years)	34	30	31	NS
% Female	57	50	53	NS
% Divorced, widowed, or separated	22	23	27	NS
% Hospitalized > 4 weeks	31	37	51	NS
% Index hospital = first hospital	23	20	14	NS
Age first ill, median (years)	31	33	25	NS
% Ill < 1 month at index	37	46	51	NS
% Previous suicide attempt	12	15	17	NS
Precipitating event	21	32	38	.01
Organic features	13	14	29	.01
Hallucinations	0	27	35	.000
Marked improvement				
At outcome	61	64	58	NS
With ECT	69	65	78	NS
Use of antipsychotic drugs	70	87	90	.001
Follow-up, 2+ years	40	40	42	NS
Suicide attempts	1	2	10	.02
Suicide	1	1	0	NS
Death from other cause	5	3	3	NS

age at first illness, acuteness of illness, or whether or not the index hospitalization was the first hospitalization. If one compares acuteness of onset in the manic patients with that in the unipolar depressive patients, the difference is noteworthy. Patients in all manic groups were more likely to have an acute onset (less than 1 month) before admission than those in the unipolar group.

Mood-incongruent psychotic manic patients were more likely than mood-congruent psychotic manic patients to have had precipitating events and were more likely to show memory defects or disorientation. Unlike the difference among unipolar patients, the mood-congruent psychotic manic patients and mood-incongruent psychotic manic patients were equally likely to have had hallucinations. Patients in both manic groups were equally likely to show marked improvement with ECT. By hospital discharge, the manic groups showed similar improvement regardless of treatment. During follow-up, patients in the mood-incongruent group made more suicide attempts, but no significant differences existed between the mood-incongruent group and other groups for actual suicides and deaths.

Finally, the data were appropriate for the application of a novel methodology: mood-incongruent psychotic symptoms in manic and unipolar patients strongly support the idea of calling them schizoaffective. Thus, one would be able to compare the mood-incongruent unipolar depressive patients and manic patients on variables that ordinarily discriminate between nonpsychotic unipolar patients and manic patients. These variables include earlier age of onset in manic patients, fewer episodes in unipolar patients, more unipolar women than men, more nonsuppressors on the dexamethasone suppression test in unipolar patients, and more acute onsets in manic patients. Table 3–3 compares only the mood-incongruent patients. These expectations remain true in this comparison.

Discussion

The method used in this study is radically different from earlier work. We have presented a continuum of severity in patient symptomology ranging from no psychosis to congruent psychotic symptoms to incongruent psychotic symptoms. Because this study covers a 10-year period, the sample is substantial.

The mood-incongruent psychotic unipolar patients and manic patients generally fit the concept of the RDC for schizoaffective, depressive, and manic illness. An important clinical point is that schizophrenia is more likely a chronic illness than an affective disorder. It is notable, therefore, that the mood-incongruent depressive patients are more likely than the nonpsychotic depressive patients to have been acutely ill (less than 6 months) when admitted. Likewise, few manic patients in any of the three groups were likely to be ill longer than 6 months (16%, 16%, and 15%, respectively).

The patients in this study meet DSM-III criteria for a unipolar or manic syndrome (i.e., major depression or mania). Thus, we have a group of patients who clearly have the syndrome of affective illness; but, in addition, two of the six groups show the presence of clear schizophrenic symptoms, namely mood-incongruent delusions and hallucinations.

Our findings are striking. The three groups do not differ substantially from each other. Nonpsychotic unipolar patients, mood-congruent psychotic unipolar patients, and mood-incongruent psychotic unipolar patients are very similar in terms of treatment response and follow-up. The same is true of

Table 3–3. Patterns in unipolar patients versus bipolar patients, both with mood-incongruent psychotic symptoms

Criterion	Patients	
	Unipolar	Bipolar
N	60	8
Age at admission, median	34	31
Age first ill, median	29	25
% female	73	53
% Index hospitalization = first hospitalization	25	14
% 4–7+ hospitalizations	26	42
% Ill < 4 weeks at admission	12	51
% Previous suicide attempts	35	17
% Organic features at index	19	29
% Receiving ECT	50	10
% Marked improvement with ECT	70	67
% Receiving lithium	18	89
% Improvement or marked improvement on lithium	55	65
% Nonsuppression on dex test		
At 8 A.M. (≥ 5 mg/dl)	28	9
At 4 A.M.	53	56
% Outcome		
Marked improvement	57	58
None or partial improvement	15	8
% Follow-up, 2+ years	41	42

the three groups in the manic illnesses. Thus, by separating the three groups, we have shown the practicality of considering them together in terms of management.

What is the difference between and among the three groupings? They are obviously different clinically, and one might consider them to be on a continuum of severity. Certainly the presence of hallucinations in the mood-incongruent psychotic unipolar patient group supports that possibility. Our major finding is that these groups look rather similar in prehospitalization background, treatment results, and follow-up. In this study, it does not make any difference, in terms of response to therapy and future course, that some patients have psychotic symptoms and others do not. Those patients with mood-congruent psychotic symptoms may, in fact, be the most homogeneous group among the depressive patients. They fit the concept of "endogenous" depression best of all. The depressive group with no psychotic symptoms probably contains a mixture of some neurotic depressive patients and some endogenous depressive patients, and it is entirely possible that the groups of acute onset incongruent psychotic manic patients and depressive patients contain a small number of patients who are schizophrenic.

On a practical basis, these illnesses are all similar. Patients with these illnesses respond to treatment well, and they have the same quality of follow-up. Of course, as was noted earlier, the diagnosis of an schizoaffective disorder depends on the criteria used. If the criteria were changed, it is quite conceivable that the results of our study would be different. That there will be a change in criteria seems probable, because somebody always seems willing to publish or pay for a change in criteria, even if no new data dictate the necessity for this.

Conclusions

We compared a group of unipolar depressive patients who had either nonpsychotic, mood-congruent psychotic, or mood-incongruent psychotic symptoms. We then did similar comparisons with a set of manic patients. These separations generally conform to some conceptions of schizoaffective disorder. The acuteness of onset in the three groups of unipolar patients and three groups of manic patients is approximately the same. The groups responded equally well to treatment within the manic and unipolar comparisons, and the follow-up was very similar. These similarities suggests that, for practical management purposes, the groups should be considered together. One finding has diagnostic significance: if one simply compares the mood-incongruent unipolar patients to the mood-incongruent manic patients, they have the same pattern of differences one sees in the nonpsychotic unipolar versus manic comparisons. The mood-incongruent manic patients are younger at onset, have more episodes, have a more acute onset, and are less likely to be women. These findings strongly suggest that mood-incongruent affective psychoses (schizoaffective disorder, manic or depressive type) are mainly related to ordinary affective disorders.

References

American Psychiatric Association: Diagnostic and Statistical Manual of Mental Disorders, 3rd Edition, Revised. Washington, DC, American Psychiatric Association, 1987

Black D, Winokur G, Nasrallah A: The treatment of depression: electroconvulsive therapy vs antidepressants, a naturalistic evaluation of 1,495 patients. Compr Psychiatry 28:169–182, 1987a

Black D, Winokur G, Nasrallah A: Treatment of mania: a naturalistic study of electroconvulsive therapy versus lithium in 438 patients. J Clin Psychiatry 48:132–139, 1987b

Levinson D, Levitt M: Schizoaffective mania reconsidered. Am J Psychiatry 144:415–425, 1987

Marneros A, Tsuang M: Schizoaffective Psychoses. Berlin-Heidelberg, Springer-Verlag, 1986

Spitzer R, Endicott J, Robins E: Research Diagnostic Criteria, 3rd Edition. New York, Biometrics Research, New York State Psychiatric Institute, 1978

Winokur G: Psychoses in bipolar and unipolar affective illness with special reference to schizoaffective disorder. Br J Psychiatry 145:236–242, 1984

Winokur G: The validity of neurotic-reactive depression. Arch Gen Psychiatry 42:1116–1122, 1985

Winokur G, Kadrmas A, Crowe R: Schizoaffective mania, family history, and clinical characteristics, in Schizoaffective Psychoses. Edited by Marneros A, Tsuang M. Berlin-Heidelberg, Springer-Verlag, 1986

Winokur G, Black D, Nasrallah A: Neurotic depression: a diagnosis based on pre-existing characteristics. Eur Arch Psychiatry Neurol Sci 236:343–348, 1987

Winokur G, Black D, Nasrallah A: Depression secondary to other psychiatric disorders and medical illnesses. Am J Psychiatry 145:233–237, 1988

4

A New Path to the Genetics of Schizophrenia

Philip S. Holzman, Ph.D.

In this chapter, I present a discussion of the genetics of schizophrenia, incorporating a historical review of genetic advances over the past 25 years as well as a look at what awaits us. New strategies in the genetics of schizophrenia are straining to be employed but are being held back by technical and conceptual restraints, which I will try to address.

Studies of major psychopathological conditions can now document accelerated progress toward solving these conditions' etiologic and pathogenetic mysteries. Evidence of this advance is found most dramatically in the unraveling of the genetic puzzles of Huntington's disease (Gusella et al. 1983), where new molecular genetic techniques have been employed. Although unlocking the secrets of schizophrenia is closer now than at any other time since Kraepelin separated that set of illnesses from the affective disorders, no clear "grand opening" can yet be celebrated. Although some of the reasons for lesser progress in schizophrenia can be attributed to the nature of the disorder—its apparent heterogeneity, its puzzling course, its peculiar familial aggregation—a principal reason may lie in the research strategy employed to solve the genetic puzzle of schizophrenia. In this chapter I will briefly review the conventional methods used to study the genetics of schizophrenia and show where they fall short of being able to exploit new molecular genetic tools. I will also present a new strategy for genetic studies that holds promise for solving the mystery of schizophrenia.

Conventional Genetic Approaches

The first phase in the study of the genetics of schizophrenia began about 1916 and continued until very recently. It addressed the question of whether schizophrenia is a heritable disorder. Family aggregation studies, twin concordance studies, and adoption and cross-fostering studies have all been used with similar conclusions: with respect to schizophrenia, something is indeed heritable. Family aggregation studies have reported mean morbid risks to siblings of schizophrenic patients of about 8% and to parents of schizophrenic patients of about 5%. Risks to offspring of one schizophrenic parent are about 12% and to the offspring of two schizophrenic parents about 40%. With respect to the rates in second-degree relatives, who share about 25% of their genetic material with a schizophrenic person, the risk is about 2% (Gottesman and Shields 1982). The data are thus consistent with an interpretation that something is inherited, but it is by no means an airtight case.

The twin method, in which concordance rates in monozygotic (MZ) and dizygotic (DZ) twins are compared, has produced data that are quite consistent with the familial aggregation method. Concordance rates among MZ and DZ sets average around 45% and 9%, respectively (Gottesman and Shields 1972; Kallmann 1946; Kringlen 1967). These rates support the assumption that schizophrenia is in some way genetically transmitted, but the high MZ/DZ ratio of 5:1 and the comparatively low concordance within MZ sets (for a fully penetrant Mendelian trait, the concordance should be close to 100%) suggest that a purely Mendelian mode of transmission is hard to maintain.

Similarly, the adoption studies confirmed that schizophrenic illness, whether in the overtly psychotic form or the milder, perhaps subclinical form (e.g., "borderline" schizophrenia, schizotypal personality disorder), runs in the biological families of schizophrenic probands but not in their adoptive families (Kety et al. 1976).

Ambiguities in the Standard Genetic Data

The rates of schizophrenia in first-degree families of schizophrenic patients are much too low to fit a model for simple Mendelian inheritance. As a consequence, geneticists have favored polygenic or multifactorial threshold models, which have been consistent with most of the data (McGue et al. 1986). The problem with such a solution, however, is that in the presence of a polygenic or multifactorial phenomenon, it becomes exceedingly difficult to conduct linkage studies and to take advantage of new molecular genetic technologies. The ultimate aim is to use these techniques to find a known marked place on a chromosome that points to an as-yet-unknown but physically close location as the site of the trait under study.

Although the data from segregation studies are quite consistent in implicating a heritable entity in schizophrenia, these studies have not been able to estimate the relative degrees of genetic and environmental influences on the presence and appearance of the phenotype. Indeed, efforts to estimate such influences have been singularly unsuccessful. Perhaps the question itself is not a fruitful one.

The questions of estimating gene-environment interaction and of fitting classical genetic models to the data base themselves on the study of the clinical phenotype, schizophrenic psychosis, about which there is more definitional ambiguity than in the study of diseases such as Duchenne's dystrophy, Becker type tardive muscular dystrophy, or Huntington's chorea, where such estimates have been quite successful. In our laboratory, instead of asking, "What is the mode of transmission?" or "What is the heritability of schizophrenia?," we prefer to ask, "What is the entity that we call 'schizophrenia' that appears to be inherited?" To answer this question requires that one reexamine the nature of schizophrenia.

An Alternative View

There is no disagreement among clinicians about defining an unambiguous case of chronic undifferentiated schizophrenia as a recombinant. But what of people who are diagnosed as "schizoaffective, depressed" or who are not at all psychotic but manifest mild, occasional thought disorder and rather weak interpersonal skills? Only the identification of the gene or genes can answer those questions, which focus on the definition of the phenotype. Yet, one must begin somewhere in the effort to find linkage. But the definition of schizophrenia is too fuzzy, and the occurrence of the unambiguous form of the disorder is too rare for linkage studies to proceed.

Consider the following: a family member of a schizophrenic patient is not psychotic, has no thought disorder, and has excellent interpersonal skills and high occupational competence. The only abnormality found in this person is a dysfunction in eye movements, an abnormality that seems to have no functional significance. Such a person, however, may indeed be a recombinant; and in this example the eye movement dysfunction (EMD) would be considered as much an expression of the "disorder" as the schizophrenic psychosis. The unit, in this example, is not the individual but the *family* that provides the arena in which the disease process is played out. The example is that of a pleiotropic disorder—that is, a process that is transmitted by a gene with high penetrance but with variable expressions of the phenotype.

Recklinghausen's neurofibromatosis is an example of such a disease. It is a phenotypically heterogeneous condition, transmitted as an autosomal dominant disorder, related to a gene located on chromosome 17, but with varying expressivity, or pleiotropy. Thus, the full syndrome occurs more rarely than do, say, two or three indicators of it. Tumors of the peripheral and cranial nerves, subnormal intelligence, café-au-lait spots, areas of skin depigmentation, and skeletal abnormalities are among the several manifestations of neurofibromatosis. It is not unusual that a family history of the disease may be overlooked or hidden in the case of a proband with the full syndrome because family members may manifest only the mildest form of a few symptoms.

As it turns out, the EMD mentioned in the example above is not a hypothetical case but rather a useful indicator of familial schizophrenia that has value as an aid in linkage studies.

Eye Movement Dysfunctions and Schizophrenia

Disorders of pursuit eye movements have been reported in about 65% of schizophrenia patients, about 40% of patients with bipolar affective disorder, and about 8% of control subjects. Patients with other psychiatric conditions, such as personality disorders or serious neurotic conditions, show a prevalence of eye movement disorders no different from that of the general population. The dysfunction occurs when the person is instructed to follow a moving object. Such following movements should be smooth and generally without interference or interruption by rapid, or saccadic, eye movements. In the impaired EMD under discussion, the eyes fail to follow the target smoothly, either because the eye moves too slowly and saccadic movements are necessary to refoveate the target, or because saccadic or other events push the fovea off of the target. Although such dysfunctions as these are usually associated with several neurological syndromes (such as Parkinson's disease, multiple sclerosis, and those accompanying hemispheric and brain stem lesions), no obvious central nervous system diseases have been reported in association with the schizophrenic and bipolar patients who show the pursuit dysfunctions.

The eye movement abnormalities under discussion here are not produced by the neuroleptic drugs that are usually prescribed for psychotic patients, although evidence suggests that lithium salts, usually prescribed for treating affective disorders, do produce pursuit abnormalities (e.g., Holzman et al. 1991; Levy et al. 1985). Central nervous system depressants, such as alcohol and barbiturates, also impair smooth pursuit eye movements. Inasmuch as schizophrenic patients with abnormal pursuit movements are able to execute rapid eye movements (saccades) with normal latency, accuracy, and speed, the pursuit dysfunction does not reflect poor attention or motivation, variables that usually interfere with the test performance of psychotic patients. Nor does chronological age account for the impairment, because most patients tested were in their 20s or 30s, a period before pursuit movements tend to degrade as a result of aging.

It is interesting to note that the same EMDs occur in about 45% of the first-degree relatives of schizophrenic patients (Holzman et al. 1984). These relatives have no history of major mental illness. Thus, the high prevalence of the pursuit disorders in the relatives of the schizophrenic patients suggests that these dysfunctions may be genetically transmitted and may represent a biological marker of a schizophrenic predisposition. This suggestion was strengthened by studies of MZ and DZ twins who are clinically discordant for schizophrenia. Despite of their clinical discordance, the twins' concordance for pursuit dysfunctions was more than 80% in the MZ sets and about 40% in the DZ sets (Holzman et al. 1980).

The specificity of pursuit dysfunctions for schizophrenia is bolstered by their frequent appearance in relatives of schizophrenic patients but not in relatives of manic-depressive patients. Studies of the relatives of patients hospitalized in Chicago and in the Boston area show that about 50% of the first-degree relatives of schizophrenic patients and about 14% of the first-degree relatives of bipolar patients manifest pursuit dysfunctions (Holzman et al. 1984). As a result, strong evidence suggests that these EMDs are associated with schizophrenia and tend to occur in families in which there is a member with clinical schizophrenia, associations that support genetic transmission of smooth pursuit tracking disorders. Their presence in a proband, however, cannot be assumed to be pathognomonic of schizophrenia, because they occur in other disorders as well.

An Apparent Discrepancy Leading to the Latent Trait Hypothesis

There is a curious, and, at first puzzling, set of facts about the association of EMDs with schizophrenia. In the Chicago-Boston sample, a number of schizophrenic patients with *normal* pursuit had healthy relatives with *impaired* pursuit. Several sets of clinically discordant DZ twins in the sample showed a similar distribution: a few schizophrenic DZ twins showed normal pursuit and their healthy co-twins showed impaired pursuit. On the face of it, this finding could not mean that schizophrenia causes impaired tracking because the healthy relative has bad pursuit. Nor could it mean that schizophrenia and EMDs are inherited together, because the schizophrenic patient in whose family the relative belongs had normal tracking.

To account for these curiosities, Steven Matthysse, Kenneth Lange, and I (1986) postulated that in many—perhaps most—schizophrenic conditions, there is a latent trait that is not directly observable but that can show itself either as schizophrenia, EMDs, or both. It is the latent trait that is to be thought of as genetically transmitted, and the transmission pattern of the latent trait may be closer to that of a single dominant gene with high penetrance than either schizophrenia or abnormal eye tracking alone. In this model, schizophrenia is the rare form of a more prevalent condition, which consists not only of schizophrenia, but also of pursuit dysfunctions and no doubt of other as yet undetermined phenotypes.

The latent trait may be thought of as a disease process in the brain that can independently invade one region or another of the central nervous system and give rise to different symptoms, depending on which region or system is affected. Thus, schizophrenia with good tracking occurs when the disease invades the less probable area and spares the more probable one. First-degree relatives will also be at risk for having the same disease process, and that process will cause EMDs (with high probability) and schizophrenia (with low probability). The two manifest traits of schizophrenia and abnormal eye movements are coupled, but it is not known why they are coupled. Perhaps such unknown but not unknowable associations are the rule in disease. For example, in Recklinghausen's neurofibromatosis, café-au-lait spots and multiple tumors of the cranial nerves are related; in phenylketonuria, light skin pigmentation and mental retardation are related. Although the coupling may seem puzzling, an understanding of the pathophysiology of the disorder solves that puzzle. In phenylketonuria, for example, the dynamics of tyrosine hydroxylase provide the required link between skin pigmentation and mental retardation.

Matthysse worked out the mathematics of the model, which was fitted to data on the distribution of schizophrenia and EMDs in families studied in Chicago and Boston. The model proposes that, although the latent trait is determined by a single major locus, it can occur without the allele that is responsible for it (as a phenocopy), and the allele can be present without the trait (as in partial penetrance). As noted above, the model further proposes that the central nervous system disease process that is the outcome of the latent trait produces clinical schizophrenia, or EMDs, or both, because it can invade one or another region of the brain independently or together. The symptoms that arise reflect the brain regions invaded.

As in most pleiotropic diseases, the full syndrome occurs more rarely than do only a few indicators of it. It is, therefore, not unusual that a family history of the disease may well be masked because only one member of the available family may manifest the full syndrome, although many others may manifest the mildest of symptoms and hence be overlooked.

In proposing this model, Matthysse, Lange, and I (1986) chose a single gene model because it is the simplest one. The equation for the model includes the following seven variables:

1. The probability of the occurrence of phenocopies;
2. The penetrance of the latent trait in heterozygotes;
3. The penetrance of the latent trait in homozygotes;
4. The probability that the latent trait gives rise to schizophrenia;
5. The probability that the latent trait gives rise to EMDs;

6. The probability of schizophrenia arising without the latent trait; and
7. The probability of EMDs arising without the latent trait.

The equation with these seven variables was then employed to search within the Chicago-Boston sample of schizophrenic and manic-depressive patients and their families for the probability of any family member having the latent trait. If an individual has the latent trait, the likelihood that he or she has either schizophrenia or EMDs, or both, can be computed using maximum likelihood mathematics for the estimates. The equation permits one to test whether the EMDs in schizophrenia, in manic-depressive illness, or in any other disease are an independent expression of the latent trait disease process or an epiphenomenon, that is, an outcome of having the disease itself.

The results of the mathematical test for the Chicago-Boston sample are as follows: In schizophrenia the data fit the latent trait model, but in manic-depressive illness the data are more easily explained as an epiphenomenon. That is, in manic-depressive illness, poor eye tracking is an outcome of the disease; in schizophrenia, it is an outcome of family transmission. Figure 4–1 reproduces the model; the left half of the figure gives the estimates of the parameters for the Chicago-Boston study.

Validation of the Latent Trait Model

In the above investigation, an equation was merely fit to the available data. The results do not represent an actual test of the model (Holzman et al. 1988). Such a test, however, was subsequently conducted using a unique sample: the *offspring* of MZ and DZ twins discordant for schizophrenia. With the collaboration of Einar Kringlen, a complete national sample of Norwegian twins was checked for individuals who had ever been hospitalized for a psychosis. After further screening, twins were selected if one and only one twin met the criteria for schizophrenia, bipolar affective disorder, or reactive psychosis. Subjects were recruited from all of Norway. Our research team traveled to test subjects in or near their homes, or the subjects came to the Psychiatric Clinic at the University of Oslo to be tested.

This study (Holzman et al. 1988) included two features that merit attention. First, it employed a different method of ascertainment from that employed in the Chicago-Boston studies. The Chicago-Boston studies began with schizophrenic and manic-depressive probands and recruited parents and siblings of those patients. Ascertainment of the Norwegian replication sample began with a total national sample of twins; from them, psychotic twins were selected, and a further selection was made of the offspring of the twins.

The second interesting feature of this Norwegian study is the factor of discordance for psychosis. One twin is affected with a psychosis, the co-twin is not. In one group, the children of an MZ twin have a biological parent who has been psychotic. In another group, the parents are apparently psychiatrically normal. Nevertheless, both parental MZ twins, regardless of their clinical psychiatric status, carry the same genes. Therefore, we expected that among the offspring of both MZ parents, regardless of their clinical histories, there should be the same prevalence of clinical schizophrenia and eye movement disorders, and the prevalence of both disorders should accord with the predictions generated from the latent trait model that was fitted to the Chicago-Boston sample.

Figure 4–1. Estimates of the parameters of the latent trait model, based on the Chicago-Boston sample in the left panel, and on the Norwegian sample in the right panel. The parameters are: 1) the penetrance of the homozygote (π_2); 2) the penetrance of the heterozygote (π_1); 3) the probability of the phenocopies (π_0); 4) the probability that the latent trait gives rise to schizophrenia (r_1^+); 5) the probability that the latent trait gives rise to eye tracking dysfunctions (r_2^+); 6) the probability that schizophrenia can arise without the latent trait (r_1^-); and 7) the probability that eye movement dysfunctions can arise without the latent trait (r_2^-).

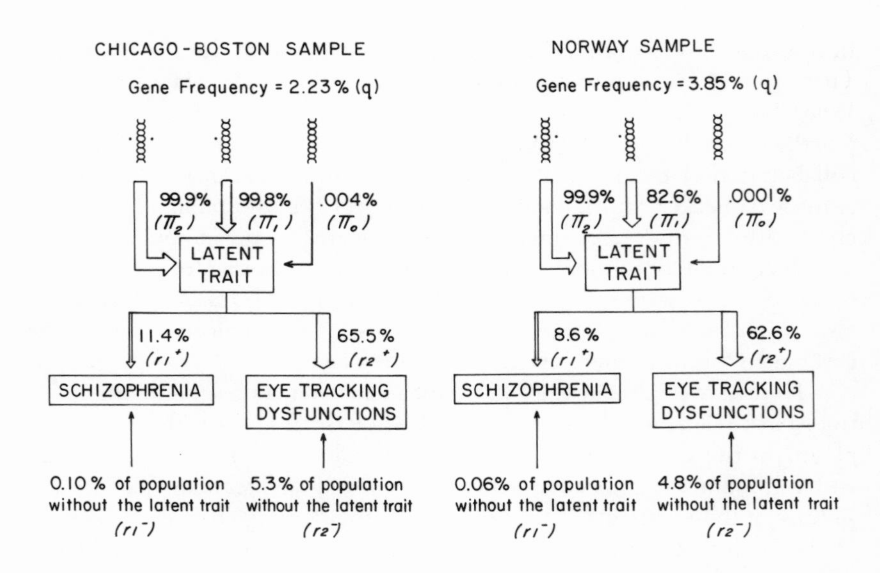

With respect to the offspring of the DZ affected twins, we expected the same prevalence of schizophrenia and EMDs among their children as among the offspring of the MZ probands and co-twins. However, among the offspring of the DZ unaffected co-twins, we expected about half the prevalence found among the offspring of probands. Among the offspring of the twins with manic-depressive psychosis or reactive psychosis and of their healthy co-twins, we expected the prevalence of EMDs to be no greater than that found in the normal population.

In this design, which tested the offspring of discordant twins, the offspring of the MZ twins provide a special kind of control over environment. Although they are actually nieces and nephews of the affected co-twin, genetically they are first-degree relatives of the schizophrenic co-twin. Thus, the study has features of an adoption study as well as of a twin study. Offspring are reared from birth in an environment that is different from that of their genetic first-degree relatives, who are legally their aunts or uncles.

The clinical diagnoses of the twins' conditions were obtained from clinical examinations performed by Kringlen over a period of 5 years in the 1960s (Kringlen 1967). His diagnostic protocol was supplemented by hospital records. The diagnoses followed Scandinavian conventions of strict criteria for schizophrenia, manic-depressive illness, and reactive psychosis; and it seems probable that the diagnostic criteria used do not differ significantly from current DSM-III-R (American Psychiatric Association 1987) criteria. We tested 213 subjects, of whom 170 were offspring of the twins, 29 were offspring of offspring, and 14 were spouses, along with some probands (there were also some subjects who bore no relationship to the study who were entered to protect the single-blind design of the study).

Pursuit eye movements were recorded without knowledge of parental diagnosis and the proband's twin status. Recording was by infrared reflected light. The signals were stored in a portable computer for later scoring and evaluation. A description of the apparatus, recording method, scoring techniques, and other technical details of the study have been reported elsewhere (Holzman et al. 1988).

Table 4–1 shows the predicted and observed frequencies of EMDs and of clinical schizophrenia for the Norwegian sample of offspring. Omitted from this table are the frequencies for the 29 children of those offspring (i.e., the grandchildren of the probands and co-twins). The predictions were based on the latent trait model generated from the Chicago-Boston sample. Three of 77 offspring of all schizophrenic probands and of their MZ co-twins displayed overt clinical schizophrenia. This prevalence of 3.9% is very close to the 3.79% predicted by the latent trait model and to the 3.7% obtained by Tsuang and colleagues (1980).

The comparison of predicted and obtained EMDs also shows noteworthy congruence (7.20 predicted and 8 obtained for the offspring of schizophrenic

MZ probands; 10.23 predicted and 10 obtained for unaffected MZ co-twins). Similarly striking are the rates of EMDs in offspring of DZ twins (6.25 predicted and 4 obtained for the DZ probands; 8.10 predicted and 9 obtained for the DZ co-twins). The chi-square analysis (χ^2 = 1.45, df = 5) indicates a nonsignificant difference between the obtained and predicted prevalences. Likewise, within each group there was no significant chi-square between the predicted and obtained frequencies of EMDs in the offspring of probands compared with the offspring of the unaffected co-twins.

Figure 4–1 shows the maximum likelihood estimates for the seven parameters obtained for the Chicago-Boston sample and for the Norwegian replication sample. The estimates are close in both samples except for the estimation of the percentage of homozygotes. When one applies the estimates of the Chicago-Boston sample to the Norwegian data, no significant difference emerges between their likelihood and the maximum likelihood (χ^2 = 0.65, NS).

Table 4–1. Predicted and observed frequencies of clinical schizophrenia and EMDs in a sample of offspring of discordant twins

	Subjects	Families	Schizophrenia		EMDs	
Parents	(#)	(#)	Predicted	Observed	Predicted	Observed
MZ proband with schizophrenia	24	10	1.18	2	7.20	8
MZ co-twin (unaffected)	33	17	1.62	1	10.23	10
DZ proband with schizophrenia	20	9	0.99	0	6.25	4
DZ co-twin (unaffected)	43	11	1.13	0	8.10	9
MZ proband with affective disorder or reactive psychosis	25	10	—	0	2.08	3
MZ co-twin (unaffected)	16	7	—	0	1.28	1

Note. MZ = monozygotic; DZ = dizygotic.
Source. Holzman et al. 1988.

When one equates the number of heterozygotes with the number of homozygotes, as required in a strict dominant model, the parameters fit the Norwegian sample with no significant chi-square. No recessive model fits the data well; and, if one excludes genetic transmission, the fit is extremely poor. The fit for the manic-depressive and reactive psychosis group is that for an "epiphenominal" model (i.e., one that is consistent with EMDs as an outcome of the psychosis or of the treatment for the psychosis).

Figure 4–2 shows a family pedigree from the Norwegian sample. It is noteworthy in this example that only one of nine offspring is schizophrenic by clinical examination. It is not surprising, therefore, that schizophrenia should appear to be only weakly familial and that there will be meager encouragement for discerning known genetic models in that familial distribution if one regards the phenotypic appearance of schizophrenic psychosis as the trait that is transmitted. Perhaps the inclusion of schizophrenia spectrum conditions would increase the prevalence of clinical phenotypes to the point of fitting a known genetic transmission pattern. But when one includes EMDs *and* clinical schizophrenia among the phenotypes that express the condition, the distribution presents a more optimistic picture for genetic modeling. In the family presented in Figure 4–2, five of the eight offspring of the

Figure 4–2. A sample pedigree from the Norwegian sample. Schizophrenia is designated by diagonal hatch marks. Eye tracking dysfunctions are designated by filled circles or squares. Proband is indicated by an arrow.

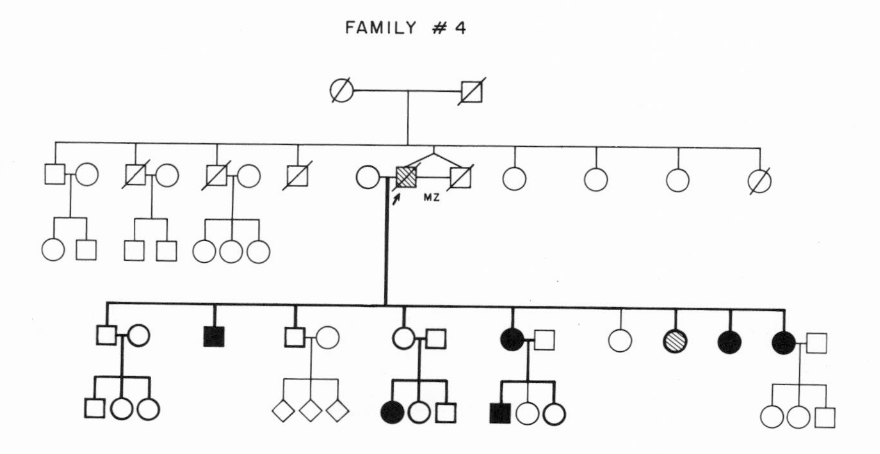

proband (the ninth offspring was unavailable for examination) show one of the two traits. Furthermore, one can see the appearance of the EMDs in the grandchildren of the proband, as expected.

When schizophrenia and EMDs are considered *together* as two independent manifestations of a latent trait, the data are consistent with the latent trait being genetically transmitted as an autosomal dominant gene with high penetrance. The transmission of the latent trait is closer to that of a classic Mendelian gene than either of the manifest traits alone. Having the gene and having the latent trait are highly associated. Although the latent trait can express itself as either schizophrenia or as EMDs, the parameters from the Norwegian sample suggest that EMDs are about seven times more likely. Using Bayesian algebraic methods, one can calculate that the latent trait has a population prevalence of 6.3% and that 89.2% of people with schizophrenia have it, as well as 49.0% of people who show EMDs. Thus, the probability of having the latent trait is increased 7.8 times if one has EMDs and 14.3 times if one has schizophrenia. With this model, one can calculate the probability that a particular family member has the latent trait. Figure 4–3 shows those probabilities for the family members represented in Figure 4–2.

Note that some nonschizophrenic members with normal eye tracking movements may have a higher probability of possessing the latent trait than the general population prevalence, which in this sample was estimated at

Figure 4–3. The sample pedigree, shown in Figure 4–2, here with Bayesian probabilities for each family member having the latent trait.

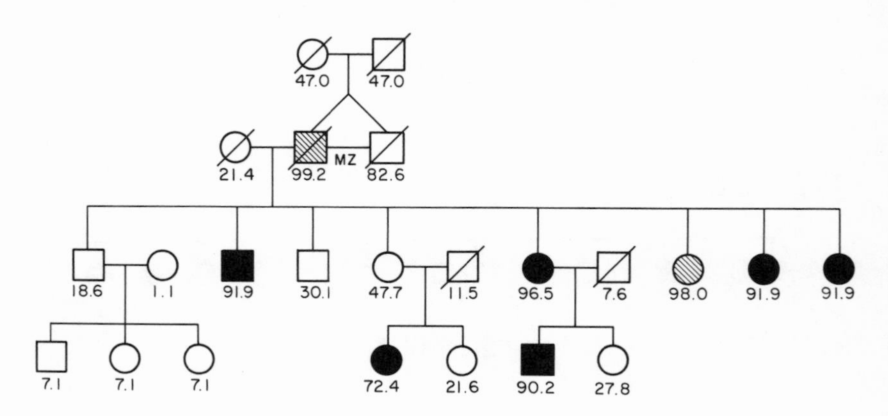

6.3%. For example, in the family illustrated here, the asymptomatic daughter of the proband has a higher probability of having the latent trait (47.7%) than does one of her asymptomatic brothers (18.6%), because one of her children has EMDs and none of the brother's three children has EMDs.

By using this method of estimating the probabilities of having the latent trait, one can raise the productive yield of molecular genetic exploration. The efficiency of efforts using techniques such as restriction fragment length polymorphisms, which rests on the assumption of major loci, depends on the accuracy of identifying the recombinants. Therefore, the calculations of latent trait probabilities illustrated here may increase the likelihood of effective molecular genetic exploration.

Some Ambiguities in the Model

A problem is presented by the latent trait model when one considers that it underestimates both the MZ concordance for schizophrenia and the recurrence risk to offspring of matings of two schizophrenic parents. For example, in this sample of MZ twins, Kringlen (1967) had determined the concordance rate to be 25% when a strict set of criteria for schizophrenia was used. However, the estimated concordance of schizophrenia in the MZ sets, which is based on the best fitting latent trait parameters, is only 6.4%. Matthysse and Kidd (1976) had pointed out that such a discrepancy is characteristic of single-gene models.

The prediction of MZ concordance and of recurrence risk in offspring of dual matings can, however, be raised by a simple modification of the latent trait model. By merely allowing the "path" from the latent trait to the manifest trait of schizophrenia to be partially genetic or even to be under some environmental control, one will substantially raise the predicted rates. This increase occurs because the MZ twins share both the genes leading to the latent trait and the genes (and most probably the environmental factor, say a virus) influencing the schizophrenia path. But such modifying genes may not be very informative about schizophrenia because their influence may be on general adaptive traits, such as interpersonal aversiveness or general intellectual features. These features may make the difference between schizophrenia and schizotypal personality disorder that may elude clinical detection, but they are likely to be polygenic and therefore unsuitable for linkage analysis. Any inherited trait that decreases coping ability, for example, would be shared by MZ twins and would likely be shared by siblings of two schizophrenic parents. But these genes would tell us very little about schizophrenia itself.

However, adding parameters to the model will necessarily exact a price in terms of power, because the more adjustable the parameters in the model, the higher the lod score (i.e., the logarithm of the odds that linkage exists) must be to reject false positives at a fixed level of confidence. The model is, after all, a heuristic device to aid in the search for linkage. If it is inconsistent with observed data, the model must be rejected. If the empirical fit is simply imperfect and the discrepancies can be corrected merely by adding parameters, our position is to add them if the corrected model increases the power to detect linkage.

A Possible Misunderstanding About the Latent Trait Hypothesis

Two studies have reported linkage between a section of chromosome 5 and schizophrenia (Bassett et al. 1988; Sherrington et al. 1988). The first report (Bassett et al. 1988), a cytogenetic study, described a family in which a psychotic proband and one of his four maternal uncles had both a schizophrenic condition and a partial trisomy of chromosome 5. All other members of this family of eight individuals were apparently normal, except for the mother of the proband, who had a balanced translocation (i.e., genetic material deleted from chromosome 5 was inserted into chromosome 1). Thus, on karyotyping, an abnormality of chromosome 5 was detected; and the two people showing this abnormality had conditions that were diagnosed as schizophrenic. Dysmorphic features were associated with the chromosomal abnormality. Because of the confluence of these conditions—a chromosome 5 trisomy, schizophrenia, and dysmorphic features—it was tempting and reasonable to begin a search for linkage between schizophrenia and markers on chromosome 5 in the area of the trisomy, 5q11-5q13.

Such a family also presents an opportunity to explore the role of EMDs in the distribution of these several abnormalities. Iacono and colleagues (1988) undertook such an exploration and reported that the proband and his uncle, mother, brother, and father had been tested. However, the three unaffected uncles of the proband had not been tested. The report stated that both the proband and his schizophrenic uncle manifested clearly abnormal pursuit movements but that the other members of the family who had been tested had shown normal smooth pursuit eye movements. They interpreted these data as supporting the latent trait hypothesis.

With some significant reservations, Matthysse, Levy, and I agreed with that conclusion, and we wrote of our concerns with respect to the use of the latent trait hypothesis in this particular family study (Holzman et al. 1989). In the Iacono report (Iacono et al. 1988), the reader is left with the impression that the latent trait hypothesis predicts a one-to-one occurrence of schizophrenia and EMDs. But an examination of Figure 4–2 shows that these two phenotypes did *not* occur jointly in the same individuals. Indeed, the theory predicts that, in large pedigrees, there should be some segregation of the two traits away from each other. It was that very observation that led us to conclude that schizophrenia and pursuit dysfunctions were independent and alternative expressions of a latent trait that may occur together or separately.

Had it been the case that EMDs occurred in some of the nonschizophrenic, nontrisomic members of this interesting family, the latent trait hypotheses would not have been disconfirmed, although a consistent finding of disjunction between the two traits throughout large pedigrees would be negative evidence. Indeed, a finding of EMDs in one or more nonschizophrenic family members would strengthen our impression that the psychiatric pathology observed in this family, despite its unusual origin and association with dysmorphology and chromosomal abnormality, is similar to the schizophrenias that we have observed in our studies. It is the appearance of eye movement abnormalities in the apparently unaffected biological relatives that distinguishes the role of those abnormalities in schizophrenia from that in other neurological, toxic, and psychiatric conditions, such as Parkinson's disease, barbiturate intoxication, or manic psychosis, where these dysfunctions may occur in probands but are uncommon in their biological relatives.

In these explorations, a conservative approach is prudent. One can agree that because the chromosome 5 trisomy occurred in the only two schizophrenic members of this family, it is not unreasonable to assume that genetic material in the trisomic section of chromosome 5 "carried" schizophrenia. However, it does remain possible that the association was only the result of chance, and the null hypothesis must be tested. A family with an uncle and a nephew with schizophrenia is not unusual. What may be unusual is that, in this family, a particular nephew (1 out of 2) and his uncle (1 out of 4) had schizophrenia. The probability of this specific outcome is $.50 \times .25 = 0.125$, which is rare enough to be of interest but not rare enough to warrant discarding the hypothesis that this is merely a chance relationship.

If the association were due to chance, abnormal tracking in some nontrisomic, nonschizophrenic relatives would be expected according to the latent trait model. Inasmuch as the proband's three other uncles were not schizophrenic and had a normal karyotype, eye movement testing in those relatives would be highly informative about whether a latent trait model with a gene located on the trisomic section of chromosome 5 can still be maintained for this family. If some of the nonschizophrenic, nontrisomic relatives show pursuit dysfunctions, the data would be consistent with the latent trait hypothesis but would also weaken the hypothesis that the latent trait gene is located on the trisomic section of chromosome 5, because relatives who are neither trisomic nor schizophrenic have pursuit dysfunctions.

It is of interest that although one study did report linkage between schizophrenia and the trisomic section of chromosome 5 (Sherrington et al. 1988), at least five other studies failed to find linkage, using the same probe

Figure 4–4. A sample pedigree from an ongoing study utilizing the latent structure model for establishing linkage. Proband is indicated by an arrow.

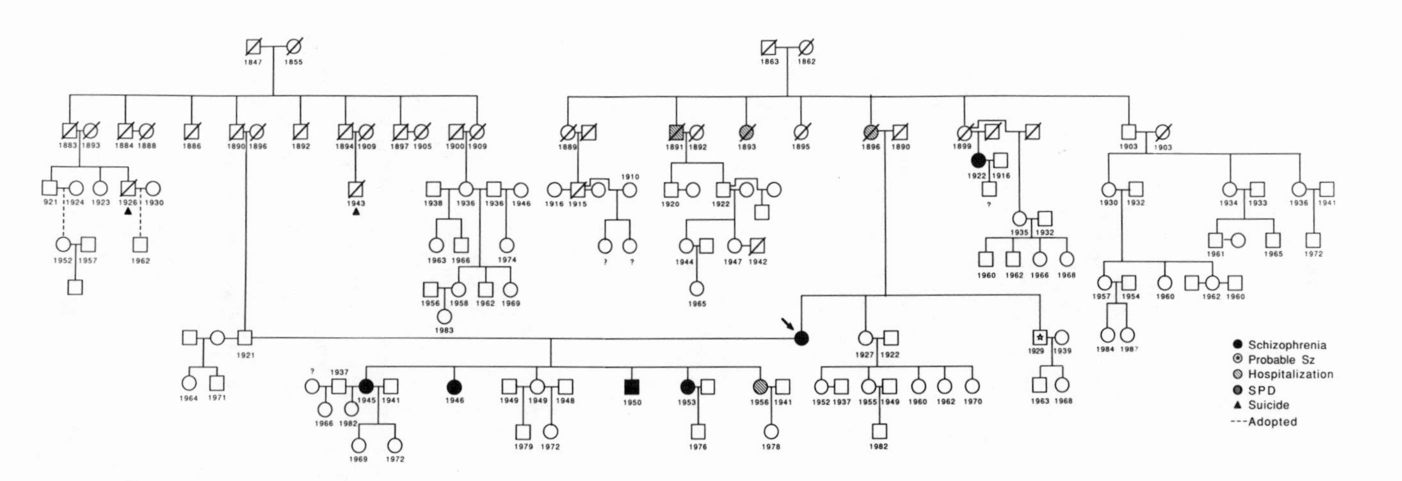

as well as others (Crow et al. 1991; Detera-Wadleigh et al, 1990; Kennedy et al. 1988; McGuffin et al. 1990; St. Clair et al. 1989). Although it is possible to regard these failures to replicate as evidence for the heterogeneity of schizophrenia, it is also quite possible that the report of Sherrington and colleagues was in error and that the confluence of trisomy, schizophrenia, and EMDs in this family is a chance occurrence.

Prospicience

The heuristic advantage of the latent trait hypothesis can be demonstrated by estimating the power of the latent structure model for establishing linkage in large pedigrees. One such pedigree is illustrated in Figure 4–4. This 118member family with five schizophrenic members is typical of the several we are currently studying. Matthysse has computed a power calculation based on a similar family. Assuming the validity of the latent trait model with the parameters as estimated in our previously published work (Holzman et al. 1988) and as republished in Figure 4–1 in this chapter, Matthysse computed the power of obtaining linkage based on a recombination fraction of 0.05. The simulation method is reported in detail by Lange and Matthysse (in press). The simulation assumed a marker locus with a large number of alleles and with all marker genotypes distinguishable from each other.

Twenty pedigrees were simulated, and the results clearly showed that the calculated power of conventional linkage analysis (using as the phenotype only clinical schizophrenia as described in DSM-III-R) is considerably lower than that when EMDs are included as an alternative phenotype. Lod scores of considerably less than 1.0 are predicted using schizophrenia alone, whereas simulations with the latent structure model give lod scores that were higher than 4.0 in 50% of the cases. This is an understandable result, because the conventional method using schizophrenia alone provides many fewer informative individuals than does the latent structure method.

I hope this chapter has succeeded in presenting a new strategy that can help to penetrate beyond our present frontier. It appears to us that the era of twin and adoption studies has powerfully demonstrated the heritability of an entity that is called schizophrenia. Now we must try to use new methods to take us beyond the frontier.

References

American Psychiatric Association: Diagnostic and Statistical Manual of Mental Disorders, 3rd Edition, Revised. Washington, DC, American Psychiatric Association, 1987

Bassett AS, McGillivray BC, Jones BD, et al: Partial trisomy of chromosome 5 co-segregating with schizophrenia. Lancet 799–801, 1988

Crow RR, Black DW, Wesner R, et al: Lack of linkage to chromosome 5q11-q13 markers in six schizophrenia pedigrees. Arch Gen Psychiatry 48: 357–361, 1991

Detera-Wadleigh SD, Goldin LR, Sherrington R, et al: Exclusion of linkage to 5q11-13 in families with schizophrenic and other psychotic disorders. Nature 340:391–393, 1990

Gottesman II, Shields J: Schizophrenia and Genetics: A Twin Study Vantage Point. New York, Academic, 1972

Gottesman II, Shields J: Schizophrenia: The Epigenetic Puzzle. Cambridge, England, Cambridge University Press, 1982

Gusella JF, Wexler NS, Conneally PM, et al: A polymorphic DNA marker genetically linked to Huntington's Disease. Nature 306:234–238, 1983

Holzman PS, Kringlen E, Levy DL, et al: Deviant eye tracking in twins discordant for psychosis: a replication. Arch Gen Psychiatry 37:627–631, 1980

Holzman PS, Solomon CM, Levin S, et al: Pursuit eye movement dysfunctions in schizophrenia: Family evidence for specificity. Arch Gen Psychiatry 41:136–139, 1984

Holzman PS, Kringlen E, Matthysse S, et al: A single dominant gene can account for eye tracking dysfunctions and schizophrenia in offspring of discordant twins. Arch Gen Psychiatry 45:641–647, 1988

Holzman PS, Matthysse S, Levy DL: Eye tracking dysfunction is associated with partial trisomy of chromosome 5 and schizophrenia: A response. Arch Gen Psychiatry 46:756–757, 1989

Holzman PS, O'Brian C, Waternaux C: Effects of lithium treatment on eye movements. Biol Psychiatry 29:1001–1015, 1991

Iacono WG, Bassett AS, Jones BD: Eye tracking dysfunction is associated with partial trisomy of chromosome 5 and schizophrenia. Arch Gen Psychiatry 45:1140–1141, 1988

Kallmann FJ: The genetic theory of schizophrenia: an analysis of 691 schizophrenic twin index families. Am J Psychiatry 103:309–322, 1946

Kennedy JL, Giuffra LA, Moises HW, et al: Evidence against linkage of schizophrenia to markers on chromosome 5 in a northern Swedish pedigree. Nature 336:167–170, 1988

Kety SS, Rosenthal D, Wender PH, et al: Studies based on a total sample of adopted individuals and their relatives: why they were necessary, what they demonstrated and failed to demonstrate. Schizophr Bull 2:413–428, 1976

Kringlen E: Heredity and Environment in the Functional Psychoses. Oslo, Norway, Norwegian Monographs on Medical Science, 1967

Lange K, Matthysse S: Simulation of pedigree genotypes in random walks. Am J Hum Genet 45:959–970, 1989

Levy DL, Dorus E, Shaughnessy R, et al: Pharmacological evidence for specific pursuit dysfunction to schizophrenia: lithium carbonate associated with abnormal pursuit. Arch Gen Psychiatry 42:335–341, 1985

Matthysse S, Kidd KK: Estimating the genetic contribution to schizophrenia. Am J Psychiatry 133:185–191, 1976

Matthysse S, Holzman PS, Lange K: The genetic transmission of schizophrenia: Application of Mendelian latent structure analysis to eye tracking dysfunctions in schizophrenia and affective disorder. J Psychiatr Res 20:57–65, 1986

McGue M, Gottesman II, Rao DC: The analysis of schizophrenia family data. Behav Genet 16:75–87, 1986

McGuffin R, Sargeant M, Hetti G, et al: Exclusion of a schizophrenia susceptibility gene from the chromosome 5q11-q13 region: new data and a reanalysis of previous reports. Am J Hum Genet 47:524–535, 1990

Sherrington R, Brynjolfsson J, Petursson H, et al: Localization of a susceptibility locus for schizophrenia in chromosome 5. Nature 336:164–167, 1988

St. Clair D, Blackwood D, Miner W, et al: No linkage of chromosome 5q11-q13 markers to schizophrenia in Scottish families. Nature 339:305–309, 1989

Tsuang MT, Winokur G, Crowe RR: Morbidity risks of schizophrenia and affective disorders among first degree relatives of patients with schizophrenia, mania, depression and surgical conditions. Br J Psychiatry 137:497–504, 1980

5

Studies of Neurobehavior: The Past 25 Years

Gary J. Tucker, M.D.

Neurobehavioral studies of psychopathology have changed drastically in the past 25 years. Twenty-five years ago the modal psychiatric investigation was one of looking for a "biologic marker" that could define a specific form of psychopathology. Currently, the major interest of neurobehavioral studies centers on the search for actual mechanisms of central nervous system dysfunction that create the psychopathology. To illuminate these changes, in this chapter I will focus on two areas of neurobehavioral studies in psychiatric patients: 1) neurologic findings in schizophrenia and 2) psychiatric symptoms reported in patients with seizure disorders.

In part these changes highlight the difference between "brain" and "mind." In the 1950s and 1960s, American psychiatry was preoccupied with mental phenomena—a "disembodied mind," if you will. However, many factors caused a shift to "brain" (Detre 1987). These included the advent of psychotropic medications and their effects on various neurotransmitters; the widespread use of psychedelic drugs and their attendant psychotic phenomena; and, more important, the movement of psychiatric care for seriously mentally ill patients into the general hospital. All of these factors helped to focus attention on the crucial role of the central nervous system in disturbances of behavior.

The movement of psychiatric care into general hospitals is of particular interest in that psychiatrists began to see patients who appeared to be psychotic, depressed, or anxious but whose psychopathology was clearly due to encephalitis, endocrine disturbances, seizure disorders, head trauma, or other "physical" problems. Most remarkable was that these conditions were phenomenologically indistinguishable from the conditions with which a psychiatrist traditionally dealt. Although these were not new causes of psychiatric phenomena, the psychiatrist now saw the entire evolution of the patient's illness, whereas previously much of the care of the seriously mentally ill was in state hospitals and often removed from the urban setting. Consequently, before the 1960s, most psychiatrists saw only the acute picture; and, because the patient was quickly sent away to a state hospital, psychiatrists rarely saw the subsequent evolution of the patients' illnesses and were thus seldom confronted with the accuracy of their diagnoses (Maxmen et al. 1974).

The shift in interest from "mental" function to "brain" function also emphasized the importance of neurology in the training of psychiatrists, as well as emphasizing the large border shared by the two specialties with which neither neurologists nor psychiatrists were capable of dealing. Such considerations as the relationship of abnormal electroencephalograms (EEGs) to behavior, the various movement disorders induced by neuroleptic medication, and the clear behavioral complications of seizures and head trauma all demonstrated broad areas of ignorance. The fact that the specialties remain separate also probably demonstrates the different interests of neurologists and psychiatrists: neurologists are primarily interested in diagnosis in relation to the anatomy and physiology of the central nervous system, and psychiatrists are fascinated not only by diagnosis but by abnormal behavior. Consequently, two subspecialties of neurology and psychiatry have emerged, one labeled in psychiatry as neuropsychiatry and the other, in neurology, as behavioral neurology. Many departments of psychiatry and neurology are developing divisions emphasizing these subspecialty areas (Tucker and Neppe 1988).

However, for the history student, the question is immediately raised: Why is this interest in the functioning of the central nervous system so prominent now? This question is especially pertinent when we consider that ventricular changes in schizophrenic patients have been noted for 62 years in large studies utilizing the pneumoencephalogram (Weinberger and Wyatt 1982). For more than 50 years, EEG abnormalities have been found consistently in schizophrenic patients (Pincus and Tucker 1985). For more than 25 years, there have been abnormal findings in the clinical neurologic examination of schizophrenic patients. It is only now, however, that these findings are considered as possibly etiologic and certainly indicative of central nervous system dysfunction, whereas in the past they were viewed as some type of incidental or inexplicable finding in a primarily mental disturbance. It is interesting that, basically, over the past 25 years, these findings have not been refuted. However, the interpretation of their meaning has changed, and the research is now directed at explaining the findings rather than the global symptoms of schizophrenia.

Neurologic Findings in Schizophrenic Patients

Over the past 25 years, studies of schizophrenic patients have consistently yielded findings of central nervous system dysfunction, including perinatal and developmental abnormalities, brain structure, EEG abnormalities, autonomic dysfunctions, vestibular dysfunctions, neuropsychological test deficits, and abnormal findings on clinical neurologic exams.

Perinatal Factors

It has been shown over the years that schizophrenic patients have lower IQs than control subjects and that this relationship is maintained over time. This is not true for patients with affective disorder. There are also more psychiatric symptoms in the prepsychotic history of children who develop schizophrenia than in those with affective disorder. Children who later develop schizophrenia have lower birth weights, decreased crying, poor muscle tone, and prominent vascular beds as well as increased "quiet" states. There is an inexplicable increase in perinatal deaths in the offspring of schizophrenic mothers. There are also more pregnancy complications in schizophrenic mothers (Campion and Tucker 1973; Erlenmeyer-Kimling et al. 1982; Pincus and Tucker 1985).

Brain Structure

In 1927, 8 years after Dandy (1919) developed the technique for pneumoencephalography, the first reports were published suggesting that schizophrenic patients had internal hydrocephalus (Weinberger and Wyatt 1982). Since then, more than 30 studies using pneumoencephalography in schizophrenic patients and showing increased cerebral ventricular size or other evidence of cerebral atrophy have been described. Many of these studies involved large samples of patients (Huber 1957, 195 subjects; Nagy 1963, more than 400 subjects). Correlations were made with increased ventricular size and poor prognosis but not with any treatment modalities. Some of these reports also showed no change in brain structure in these subjects, even though the clinical state changed. However, few subjects showed progressive symptomatology with increased atrophy (Weinberger 1982). In 1989, Coffman confirmed these findings by reporting that at least 40 studies using CAT scans had shown increased lateral ventricles, increased third ventricle, and sulcal widening in these patients. The implications of the cerebral atrophy and regional asymmetries were not clear. Although all of this renewed interest in the gross structure of the brain in schizophrenic patients is undoubtedly related to the ease of using the new imaging techniques, it is of note that most of the changes described are nonspecific and, when compared with neurologic disease, not as severe. Consequently, the meaning of these tantalizing findings is still unclear.

In terms of gross neuropathologic studies, the brains of schizophrenic patients have been described as being lighter (6%) (Brown et al. 1986) and having increased lateral ventricle size, particularly in the anterior and temporal horns, with thinner perihippocampal cortices. Volume reductions in the limbic temporal structure have also been noted (Bogerts et al. 1985). On microscopic investigation of the brains of schizophrenic patients, long vertical axons in the singular cortex with very few horizontal axons have been reported (Benes et al. 1987; Kovelman and Scheibel 1984).

EEG Abnormalities

Since 1964, at least 100 studies showing EEG abnormalities (usually slowing) in schizophrenic patients have been reported. The more sophisticated EEG investigations (e.g., computerized EEGs, frequency analyses) have found even higher rates of EEG abnormalities in these patients. Consequently, the rate of EEG abnormality varies from 30% in a general psychiatric population to almost 80%, depending on the sophistication of the techniques used (Pincus and Tucker 1985).

Interestingly, for a number of years, *vestibular function* has been noted to be disturbed in schizophrenic patients. Fifteen studies from different laboratories have found decreased nystagmus responses to galvanic, caloric, or rotational stimuli. This is true in both medicated and unmedicated patients. These changes are different from what Holzman and colleagues (1988) have noted in the saccadic eye movements of schizophrenic patients (Pincus and Tucker 1985).

Perhaps the most consistent findings over the years have been changes in the *autonomic nervous system*. Increased arousal, as manifested by heart rate changes, has been described consistently in schizophrenic patients. Also, increased galvanic skin responses, poor habituation, and increased reaction times have led to the development of theories about the inability of people with schizophrenia to change their mental set (Pincus and Tucker 1985).

When schizophrenic patients have been tested with *neuropsychological batteries*, it has been consistently noted that these batteries, usually used to differentiate organic from nonorganic problems, have been unable to differentiate chronic schizophrenic patients from organic schizophrenic patients. This finding has been interpreted to mean that these batteries are not valid measures for schizophrenic patients rather than that the tests are, perhaps, picking up some type of central nervous system dysfunction in these patients. Interestingly, many of the tests used to distinguish thought disorder in schizophrenic patients, such as the object sorting test developed by Kurt Goldstein, were developed to assess brain damage (Goldstein and Scheerer 1941); and the general interpretation now is that these tests do reflect central nervous system dysfunction (Pincus and Tucker1985).

Although the neurologic exam has been basically free of major abnormal findings in schizophrenic patients, minor clinical findings ("soft signs") are consistent. Soft signs represent diffuse nonlocalizing dysfunctions of the central nervous system as opposed to "hard signs," which connote specific anatomical locations (Table 5–1). As noted in the table, there is tremendous potential variability in these studies, because the number of potential soft signs is large and the standardization for replication is poor. A recent review by Heinrichs and Buchanan (1988) cited more than 30 studies of these neurologic findings in schizophrenic patients. Fifty percent to 60% of the schizophrenic patients in the studies reviewed showed increased soft signs, whereas only 5% of the controls showed such signs. The implications of these findings are that the soft signs in schizophrenic patients indicate difficulty in the area of *sensory integration*, as manifested by deficiencies in bilateral extinction, audiovisual integration, graphesthesia, and stereognosis, although there was no defect in these patients' sensory systems. Second, some degree of *motor coordination difficulties* was evident as represented by the presence of intention tremor, balance and gait dysfunctions, an inability to hop consistently, poor finger-thumb opposition, and dysdiadochokinesia. Strength was intact as well as deep tendon reflexes. The third area of dysfunction was in the general area of *decreased sequencing of movements or patterning*.

Correlation between soft signs and CAT scan abnormalities, paroxysmal EEGs, and severity of measures of thought disorder is minimal. Although there is some correlation with chronicity, there is none with premorbid social

Table 5–1. Typical parameters studied as soft signs

Stereognosis	Auditory visual integration
Graphesthesia	Vestibular dysfunctions (decreased or increased responses)
Bilateral simultaneous stimulation	Cranial nerve disorders that are nonlocalizing
Coordination, balance, gait	Diffuse EEG abnormalities
Sensory (light touch, position, vibration, pain, slow extinction, or poor lateralization)	Decreased or increased reflexes
Synkinesis	Hoffman's sign
Choreoathetosis	Poor clonus
Dystonia	Mental retardation
Tremor, tics	Memory disturbances
Speech disturbances	Nonlocalizing indicators of cerebral dysfunction on various psychological tests
Adventitious motor overflow	
Motor impersistence	

functioning. Nor is there clear correlation with positive or negative symptomatology or clear family distribution. There is no correlation with age in the schizophrenic patients, and most of the researchers believed that these findings are unrelated to medications. In general, the presence of soft signs in schizophrenic patients seems to represent a stable finding.

At best, these well-described neurologic findings in schizophrenic patients represent a strange and theoretically unrelated collection of findings that has added little to our clinical understanding of the disease. Future studies of neurologic findings must use consistently replicable techniques and correlate these findings with age, sex, medication, and stability over time. Certainly, correlations with the newer dynamic techniques for investigating the central nervous system, such as magnetic resonance imaging (MRI) and positron-emission tomography (PET), will be important. Although all of these studies lend little toward a comprehensive understanding of schizophrenia, they have laid the groundwork for investigators to look at mechanisms that may explain this still mysterious disorder. One promising area is the examination of obstetrical complications as a potential causal factor of many of these neurologic findings in schizophrenic patients. Another promising area is the application of highly sophisticated technology to examine saccadic eye movements in both schizophrenic patients and nonschizophrenic control subjects. Both of these areas have been made possible by the advent of modern technology and its application to these very old problems.

Obstetrical Complications and Schizophrenia

Schizophrenic patients have two times the obstetrical complications of other psychiatric patients (17% versus 8%) (Lewis and Murray 1987). When children who have a history of obstetrical complications have been investigated using new imaging techniques, the obstetrical complications surrounding their birth have been found to be highly predictive of increased ventricular size as well as increased width of cortical sulci. Murray and other researchers, in an interesting series of papers reviewing both the obstetrical complication literature and recent data on the development of the central nervous system, have postulated a relationship between obstetrical injury and the development of schizophrenia (Lewis et al. 1989; Mednick et al. 1989; Murray et al. 1988). They postulate that the obstetrical complications can lead to many of the diverse neurologic findings present in schizophrenic patients. Murray and colleagues go on to speculate that, although the neurologic abnormalities are the same for all patients who have had obstetrical complications, it is this insult coupled with the appropriate genetic predisposition that leads to the development of schizophrenia.

Apparently, the blood vessels in the neonatal brain are extremely sensitive to hypoxia and hypercapnia. When this secondary vascular damage occurs during the neonatal period, hemorrhage occurs in the frontal, temporal, and occipital horns, which in turn leads to periventricular necrosis with resulting enlargement of the third and lateral ventricles. Interestingly, these are the changes that have been described in schizophrenic patients. These vascular changes can have profound effects. For example, in newborn rats exposed to anoxia, a 25% increase in β-adrenergic receptors has been observed in the hippocampus of these rats in adult life. Murray also describes two types of changes evident in these animal models. One type of damage is a resultant increased neural dysplasia that disturbs axonal migration. Another type of damage that has been observed from the same injury is not evident until maturational changes during adolescence (myelinization of areas of the hippocampus does not occur until adolescence). Consequently, there is late onset of some symptoms due to the perinatal insult. Interestingly, these late changes are apparent in the dorsal lateral prefrontal areas of the cerebral cortex, areas that have recently been implicated in schizophrenia.

Although obstetrical injury does not explain all, it certainly provides a more coherent framework than we have previously had to explain the diverse neurologic findings in schizophrenia. Certainly, the timing of the result of these injuries is of interest: soft signs and poor social development early in life and the onset of schizophrenia in adolescence. Although these speculations may explain the diversity of findings, they still do not reveal much about the mechanisms of the abnormal behavior.

Saccadic Eye Movement and Schizophrenia

Abnormalities in saccadic eye movements in schizophrenic patients are a fairly consistent finding (Holzman et al. 1988). Recently, investigators have begun to explore saccadic eye movements using precise infrared tracking mechanisms and computer analysis (Hommer and Radant 1989; Hommer et al. 1986, 1991). In the nonschizophrenic subject, using a task designed to elicit saccadic eye movements, only one type of saccade—corrective saccades—have been shown. When one examines schizophrenic patients using the same task, there are frequent maladaptive "anticipatory" saccades. The functional neuroanatomy of the anticipatory saccadic eye movements has been well worked out in animals; three brain areas are involved in this simple process: the dorsolateral prefrontal cortex, the frontal eye fields, and the posterior parietal cortex. Each of these regions analyzes some aspect of the temporal and spatial responses to visual objects. All of these regions seem to be simultaneously involved in the process of prediction of where to focus the eyes, because they all possess neurons that have been observed to fire before the saccadic eye movements take place in response to anticipated targets (Goldman-Rakic 1989). The simultaneous firing of these three areas represents a parallel processing of stimuli that seems necessary for the performance of this simple act (Figure 5–1). It is also of interest that changes in dopamine and GABA concentrations at the receptor sites have been demonstrated to produce these anticipatory saccades as well. These studies show much promise for illuminating an aspect of behavior long noted in schizophrenic patients—attentional deficits. These studies also demonstrate the complexity of functional localization in the central nervous system. Consequently, many of the studies that purport to implicate the dorsolateral prefrontal cortex as being disturbed in schizophrenic patients may be overlooking the importance of complex loops in this area (Goldberg et al. 1989). These studies begin to

Figure 5–1. The basal ganglia thalamocortical circuit (oculomotor loop). (Reprinted with permission from Hommer and Radant 1989.)

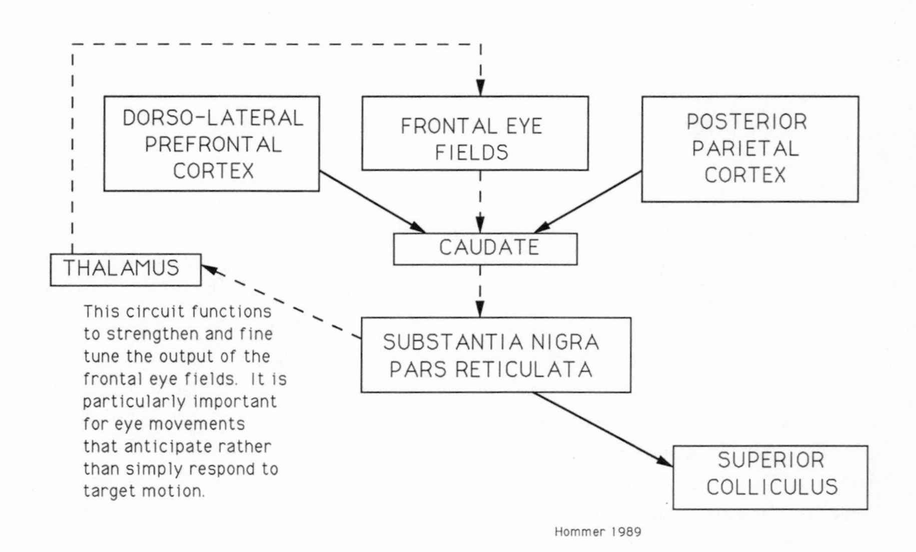

link what apparently was described as a simple marker to a possible explanation of psychopathology.

Seizure Disorders and Psychosis

In 1963 Slater, Beard, and Glithro published a series of papers describing the "Schizophrenia-Like Psychoses of Epilepsy." In these papers, they showed that all the symptoms of schizophrenia could be present in patients with temporal lobe epilepsy. They also explained that, in these same patients, there was no genetic or familial relationship to schizophrenia. Another important observation was that, although all the symptoms of schizophrenia were present, the personality of these patients was well preserved. Since that time, Trimble and others have confirmed these clinical observations and have used various standardized interview schedules to show that, at a moment in time, a patient with complex partial seizures and psychosis is phenomenologically identical to a schizophrenic patient (Perez and Trimble 1980; Trimble 1985). The incidence of psychosis in patients with seizure disorders varies from about 2.8% to 27%, with the best estimate being 7%. The occurrence of epilepsy in psychiatric populations is about 2% to 3%, and the incidence of psychosis in people with epilepsy is more than 10% over a 30-year period (McKenna et al. 1985). All of these percentages are much higher than one would expect from the incidence of schizophrenia.

Although the incidence of psychosis in patients with seizure disorders is definitely increased, the exact etiology is unclear and the studies of the relationship have been few. However, two major theories have emerged: the antagonistic theory and the affinity theory. Perhaps the most interesting is the antagonistic theory. In 1935, Meduna postulated that there was an antagonism between seizures and psychosis. Landholt, in the 1950s, described two groups of epileptic patients (Landholt 1958). In one group during psychosis, EEG records showed increased seizure activity, and, in another group during psychosis, the EEGs changed from showing seizure activity to being normal, which Landholt called "forced normalization." Pakalnis and colleagues (1987) reported seven cases in which the EEG normalized when the psychotic symptoms disappeared. This intriguing relationship has yet to be followed up; it represents one of the few hard findings of a "marker" (EEG abnormality) present in psychosis but not present when the person becomes nonpsychotic. However, the whole theory is suspect in that the vast majority of epidemiologic studies show a high comorbidity of psychosis and seizures, and the "forced normalization" group may only be a small, specific subset of these patients.

The affinity theory postulates various factors secondary to the seizures that may play a role in the causation of the psychosis (e.g., secondary to a clouded state caused by the seizures themselves; psychosocial rejection of the seizure patient leading to psychotic phenomena; various concerns about anatomic location such as specificity of the temporal lobe versus the generalized seizures; medication effects leading to decreased vitamin B_{12} and folate). Perhaps the most interesting postulate has been the relationship of the psychosis to kindling. Although kindling has been described primarily in animals, it has served as an intriguing theoretical postulate about the role that subictal stimulation of the central nervous system may play in the production of abnormal behavior (Neppe and Tucker 1988a, 1988b). Certainly, the use of anticonvulsants in many psychiatric conditions, as outlined by Post, and his speculations about the role of kindling in these behavioral disorders, has lent much interest to these speculations (Post 1988). However, none of these factors has been demonstrated as the cause of the psychotic episodes seen in seizure patients. Seizure disorders and their psychotic phenomena represent an experiment of nature that must and should be investigated with the technology now available to us, particularly in light of the fact that many of these psychotic phenomena are present episodically and would render a study population that has the presence or absence of psychosis with a reasonable periodicity. In essence, this fascinating, naturally psychotic condition has been clearly defined but has yet to be seriously investigated using modern technology.

Conclusion

Although every generation feels that it is at the threshold of new understanding, it is apparent that over the past 25 years, investigations of neurobehavioral functioning in psychopathology have changed mainly in focus from quests for markers of disease and attempts to diagnose disease to actual attempts to understand the mechanism of the psychopathology. This progress has been made possible by applying various new technological advances to the study of the central nervous system. How fruitful the application of these techniques will be will depend on the cleverness of our questions and our persistence in the face of great complexity.

References

Benes F, Majocha R, Bird E, et al: Increased vertical axon numbers in cingulate cortex of schizophrenics. Arch Gen Psychiatry 44:1017–1021, 1987

Bogerts B, Meertz E, Schonfeldt-Bausch R: Basal ganglia and limbic system pathology in schizophrenia. Arch Gen Psychiatry 42:784–790, 1985

Brown R, Colter N, Corsellis A, et al: Postmortem evidence of structural brain changes in schizophrenia. Arch Gen Psychiatry 43:36–42, 1986

Campion E, Tucker GJ: A note on twin studies, schizophrenia and neurological impairment. Arch Gen Psychiatry 29:460–464, 1973

Coffman J: Computed tomography in psychiatry, in Brain Imaging. Edited by Andreasen N. Washington, DC, American Psychiatric Press, 1989, pp 1–66

Dandy WE: Roentgenography of the brain after injection of air into the spinal cord. Ann Surg 70:397–403, 1919

Detre T: The future of psychiatry. Am J Psychiatry 144:621–625, 1987

Erlenmeyer-Kimling L, Cornblatt B, Friedman D, et al: Neurological, electrophysiological, and attentional deviation in children at risk for schizophrenia, in Schizophrenia as a Brain Disease. Edited by Henn F, Nasrallah H. Oxford, England, Oxford University Press, 1982, pp 61–98

Goldberg T, Berman K, Weinberger O: An orientation to work on the prefrontal cortex in schizophrenia, in Schizophrenia. Edited by Schulz C, Tamminga C. New York, Oxford University Press, 1989, pp 227–246

Goldman-Rakic P: Circuitry of prefrontal cortex and regulation of behavior by representational knowledge, in Handbook of Physiology, Vol 5. Edited by Plum F, Mountcastle V. New York, McGraw-Hill, 1989, pp 373–417

Goldstein K, Scheerer M: Abstract and concrete behavior. Psychological Monographs 53(2):239, 1941

Heinrichs D, Buchanan R: Significance and meaning of neurologic signs in schizophrenia. Am J Psychiatry 145:11–18, 1988

Holzman P, Kringlen E, Matthysse S, et al: A single dominant gene can account for eye tracking dysfunctions and schizophrenia in offspring of discordant twins. Arch Gen Psychiatry 45:641–647, 1988

Hommer DW, Matsuo V, Walkalwitz O, et al: Benzodiazepine sensitivity in normals. Arch Gen Psychiatry 43:542–551, 1986

Hommer DW, Radant AD: Excessive anticipatory saccadic eye movement in schizophrenia. Society for Neuroscience Abstracts 15:1205, 1989

Hommer DW, Clem J, Litman R, Pickar D: Maladaptive anticipatory saccades in schizophrenia. Biological Psychiatry 30:779–794, 1991

Huber G: Pneumoencephalographische und psychopathologische bilder bei endogen psychosen. Berlin, Springer-Verlag, 1957

Kovelman J, Scheibel A: A neurohistological correlate of schizophrenia. Biol Psychiatry 19:1601–1621, 1984

Landholt H: Serial electroencephalographic investigations during psychotic episodes in epileptic patients and during schizophrenic attacks, in Lectures on Epilepsy. Edited by Lorentzde Haas AM. Amsterdam, Elsevier, 1958, pp 91–133

Lewis S, Murray R: Obstetric complications, neurodevelopmental deviance, and risk of schizophrenia. J Psychiatr Res 21:413–421, 1987

Lewis S, Owen M, Murray R: Obstetric complications and schizophrenia, in Schizophrenia. Edited by Schulz C, Tamminga C. New York, Oxford University Press, 1989, pp 56–68

Maxmen J, Tucker G, Lebow M: Rational Hospital Psychiatry. New York, Brunner/Mazel, 1974
McKenna P, Kane J, Parrish K: Psychotic syndromes in epilepsy. Am J Psychiatry 142:893–904, 1985
Mednick S, Machan R, Huttimen M: Disturbances of fetal neural development and adult schizophrenia, in Schizophrenia. Edited by Schulz C, Tamminga C. New York, Oxford, 1989, pp 69–77
Meduna L: Versuche uber die biologische beeinflussung des ablaufes der schizophrenie. Zeitschrift für die gesamte Neurologie und Psychiatrie 152:235–262, 1935
Murray R, Lewis S, Owen M, et al: The neurodevelopmental origins of dementia praecox, in Schizophrenia. Edited by Bebbington P, McGuffin PP. London, Heinemann, 1988
Nagy K: Pneumoencephalographische befunde bei endogen psychosen. Nervenarzt 34:543–548, 1963
Neppe VM, Tucker GJ: Modern perspectives on epilepsy in relation to psychiatry: classification and evaluation. Hosp Community Psychiatry 39:263–271, 1988a
Neppe V, Tucker GJ: Modern perspectives on epilepsy in relation to psychiatry: behavioral disturbances of epilepsy. Hosp Community Psychiatry 39:389–396, 1988b
Pakalnis A, Drake M, John K et al: Forced normalization. Arch Neurol 44:289–292, 1987
Perez M, Trimble MR: Epileptic psychosis: diagnostic comparison with process schizophrenia. Br J Psychiatry 137:245–249, 1980
Pincus J, Tucker G: Behavioral Neurology, 3rd Edition. New York, Oxford University Press, 1985
Post R: Time course of clinical effects of corbamazepine. J Clin Psychiatry 49(suppl):35–46, 1988
Slater E, Beard AW, Glithro E: The schizophrenia-like psychoses of epilepsy. Br J Psychiatry 109:95–150, 1963
Trimble MR (ed): Psychopathology and Epilepsy in the Psychopharmacology of Epilepsy. Chichester, England, Wiley, 1985, pp 1–16
Tucker G, Neppe V: Neurology and psychiatry. Gen Hosp Psychiatry 10:24–33, 1988
Weinberger D, Wyatt R: Brain morphology in schizophrenia, in Schizophrenia as a Brain Disease. Edited by Henn F, Nasrallah H. New York, Oxford University Press, pp 148–175, 1982

6

Behavioral Considerations in Clinical Research and Psychiatric Medicine (Neither a Mentalist Nor a Reductionist Be!)

Joseph V. Brady, Ph.D.

The rather presumptuous title to this modest contribution honoring Thomas Detre belies the limited hands-on experience of a laboratory behavioral biologist with the clinical aspects of psychiatric research. And although I have had my nose in the psychiatry tent for the better part of the past half-century, I can hardly lay claim to any clinical expertise in the realms of depression and schizophrenia upon which this distinguished commemoration is focused. Despite being armed with a rigorous set of behavior analysis principles, buttressed by extensive laboratory experimentation, I am generally appalled when confronted with concrete clinical case problems and find myself praying for the fullest operation of beneficial nonspecific effects (Frank 1961).

Basic Concepts and Issues

As a card-carrying behaviorist, however, I do tend to favor alternatives to the dominant "inner states" orientation of psychiatry. In this regard, my views are more compatible with an environmentalism philosophy that has two main features. The first holds that knowledge comes from experience rather than divine revelation, innate ideas, or other obscure sources; and the second holds that action is governed by consequences rather than by instinct, will, attitudes, beliefs, or even the currently fashionable cognitions. Taken together, these two constructs about human conduct—the experiential basis of knowledge and the governance of action by consequences—define a philosophy of social optimism that says, if there is reason for people to be a certain way or do certain things, circumstances can be arranged. These two features of environmentalism have rather different histories in the development of philosophical thought (e.g., the writings of Hume and Darwin), but they appear to have come together in late 19th-century England, and their coalescence dates the emergence of modern behaviorism.

The influence of these concepts on medicine, as applied biological science, has developed, understandably, much more slowly amidst dominant biochemical and physiological orientations. But their impact has begun to find expression in current emphasis on environmental and behavioral factors in both the acute and chronic disorders that are of growing concern to a seriously overburdened health care delivery system. There is now even a fast growing discipline that flies under a banner labeled "behavioral medicine" (Pomerleau and Brady 1979), a confluence of terms to which my old friend Danny Freedman once responded wryly, "It sounds like something you give kids when they're bad."

The prescriptions for clinical psychiatry that flow from this behavioristic perspective are deceptively straightforward:

- Systematic analysis of the undesired behavior and its context will reveal the conditions of its maintenance.
- These circumstances can then be manipulated to make alternative desirable behaviors more probable.
- Such behaviors should be effectively rewarded and conditions arranged so that the desired behavior can continue to be supported by the normal contingencies of life, thus providing for the necessary carryover beyond the clinical setting.

The ease of this solution is obviously illusory, for several reasons. It presumes that we can identify, functionally, the behavior and the environmental events, both inside and outside the skin, that control it; that we can identify the type of control being exerted; and that we have sufficient jurisdiction over the situation to permit us to modify it in ways that will produce the desired changes. In principle, these presuppositions are fulfilled by a behavioral analysis followed by appropriate biobehavioral engineering. In practice, of course, behavioral principles tend to be generalized and abstract, whereas the laboratory technology upon which they are based is highly particularized. However, these principles can be useful to the extent that they suggest overall outlines of a solution that is certainly no more complex than the alternative focus on mental states that have neither weight, extension, nor other physical properties.

A major bone of contention in this regard continues to hinge on the somewhat divergent views of verbal behavior that characterize behavior analysis and psychiatric medicine. From the operationally positivist perspective of the behaviorist, verbal behaviors, both vocal and nonvocal, are most parsimoniously viewed as learned performances maintained by their environmental consequences, and under obvious audience control. As such, they require analysis within the same systematic framework as other instrumental performances. The contrasting view regards these verbal performances somewhat reductionistically and mentalistically as reflections of ideas, beliefs, concepts, cognitions, and the like—inner mediators that exercise powerful directive effects on overt behavior, very much as strategies determine tactics. Without underestimating the control exercised by such private events, observed only by the speaker, the focal importance of external environmental factors must be taken into account. Verbal behavior, no matter how private its subject matter may appear to be, is controlled by the external environment to a significant extent.

This fact does not in any way prejudice the case for an interest in what patients have to say about what they consider their own private experience. Verbal behavior is clearly the most convenient avenue of access to anything that might be considered a significant aspect of human knowledge, including one's own knowledge of oneself. This kind of highly personal interest should lead inevitably to a concern, at least in part, with the environmental events that have acted to teach the person to talk. Thus, a parsimonious account of complex verbal behavior and private events requires that the possibilities for analysis within the framework of principles already developed empirically on the basis of observation and experiment be exhausted before explanatory fictions are entertained.

These conceptual differences notwithstanding, it seems fair to say the positive effects of introducing the behavioral perspective into clinical research and psychiatric medicine will arise less from current technological applications than from the fundamental principles emphasizing maintenance and control of behavior by past and present environmental consequences. This emphasis should lead to closer and more detailed analysis, whether focused on single case treatment models or on more ambitious group research designs. It is hoped that these behavior analysis formulations, equivalent in important dimensions to diagnoses, can be highly specific, based on observations of what is happening in the patient's here and now. Viewed in this context, even the individual treatment setting can become a single case experiment—an approach that psychiatric medicine can then share with other applied biological science disciplines.

I am, of course, well aware of the new kid on the block—the currently fashionable *cognitive surrogate* for inner mental mediators. Under the rubric of "cognitive science," we have been regaled by claims that "revolutionary advances in our understanding of the nature of mind are on a par with understanding the evolution of the universe, the origin of life, and the nature of elementary particles!" This modest assessment appears in a report issued by the National Academy of Sciences (Estes 1983) over the signatures of the most visible luminaries in this nascent field. Moreover, the movement is explicitly characterized as revolutionary because it has succeeded in overthrowing the predominantly positivist and operational views of behavioral methodologies and philosophy of science that provided guarantees "against metaphysical and mentalistic explanations of human behavior" (Simon 1980). The result, according to an ever-growing claque of adherents, is a new sophistication, a new confidence, and a great gain in precision and rigor!

Somewhat paradoxically, this new cognitive science manifests notable regressive features in espousing the traditional view that behavior starts within the organism. We think and then act; we have ideas and then speak; we have feelings and then express emotion; we intend, decide, and choose before doing things. This linearly causal, mentalistic account is in sharp contrast to the demonstrably more effective operational appeal to antecedent events, contingent consequences, environmental context, and the reactional biographies of both individual and species.

Cognitive science makes the person the initiator by adopting the paradigm of information processing as the central activity involved in behavior. But when one considers that folks have been processing information for thousands of years by making records on clay, tiles, papyrus, vellum, paper, magnetic wire and tape, and now on silicon chips that are stored, retrieved, and responded to in a manner more or less like the originals, one is tempted to ask, "So what else is new?" Perhaps, more important, we might ask how taking these practices as a model or metaphor have enhanced our understanding of behavior. We are presumably enlightened by a cognitive account of how knowledge retrieved from storage affects not only what is seen but how readily it is seen. Familiar words, for example, are seen more readily than rare words, expected words more readily than unexpected ones, and decorous words more readily than obscene ones. But the same facts are explained much more parsimoniously (and operationally) by an account that appeals to the consequences of past environmental interactions, both appetitive and aversive.

Perhaps my most crucial concern is that the end product of the cognitivists' primrose information processing path always leads to construction of a representation with the inevitable need to posit an "inner person" who transforms all of this data into "precepts" and "rules." The evidence for this, of course, is presumed to be found in the way one comes to "know" about contingencies in the world. But the behavioral operations involved in the *communication* of this *knowledge* are invariably verbal—in the broadest sense of performances, both vocal and nonvocal, maintained by their effect on an audience—and this verbal behavior must be analyzed as such. In contrast, cognitive science finds it more appealing to speculate about internal processes that are neither observed nor observable. Among the more obvious problems raised by embracing such mentalistic concepts in either a basic or applied biological science is their definition in terms of constructions that have neither weight, extension, nor other physical properties. Metaphorical and, in some prominent instances, allegorical referents replace operational language, diverting attention from confrontable realities and encouraging appeals to explanatory fictions. Mental processes are represented as causal agents, and the view of "autonomous man" that emerges is inimical to the methods of procedure and rules of evidence that characterize the tradition of natural science, a tradition that has proven particularly fruitful in other areas of applied biology and in the general conduct of human affairs.

There are among the defectors from the more operationist ranks those who refer to themselves as *cognitive behaviorists*, a term that has begun to appear with increasing frequency in recent clinical and research publications. I can only take this puzzling reference to be a symptom of the loss of the sacred paradigm of the founding community of scholars. In a sense, one is tempted to label this the oxymoron of the decade.

Behavioral Mechanisms in Clinical Psychopathology

But on a less contentious, more constructive note, let me bring this discussion into contact with the psychiatric research focus that has characterized the distinguished career of our honored colleague, Thomas Detre, by calling attention to several more recently published reports that identify behavioral mechanisms in clinical psychopathology. For example, Salzinger and colleagues have produced a series of reports based on systematic experimental observations adducing evidence that schizophrenic behavior is controlled by stimuli that are immediate in the patient's environment (Salzinger 1973, 1980, 1984; Salzinger et al. 1966, 1970; 1978). Several other behavioral features of these studies establish that schizophrenic patients show relatively more stimulus constancy than object constancy, their behavior extinguishes more rapidly, their speech is impaired with respect to more remote relational aspects, and their verbal associations are controlled more by the topography of stimulus words ("clang associations") than their functional relations ("meaning"). These findings suggest a technical analysis of clinically characterized thought disorder and paranoia, for example, in terms that point to specific behavioral deficits in contextual control or what Blakely and Schlinger (1987) have recently called "function-altering contingency-specifying stimuli" (rule-governed behavior). A similar account may provide some enlightenment with respect to the pathognomonic "flatness" or lack of so-called "emotional responsiveness" that has characterized clinical descriptions of schizophrenic behavior since at least the time of Kraepelin.

None of this is to gainsay the obvious methodological and conceptual challenges that have enhanced opportunities for clinical research and applications related to more fundamental biological mechanisms in such disorders. Indeed, this field of inquiry has been cultivated assiduously, and the current lively interest in depression and schizophrenia is generously reflected in the rich fare contained in this book. It is unfortunately true, however, that dedication and industry, even of the most intense sort, do not always guarantee authentic scientific achievement. In applied biological science, and particularly psychiatric medicine, wide gaps frequently separate clinical and experimental operations on the one hand and interpretive formulations on the other. Progress in developing systematic and coherent conceptualizations that serve to integrate and unify interactive levels of discourse can be painfully slow. Even at the most fundamental biological level, there appear to be few generally accepted theoretical formulations that can bring conceptual order to the rapidly expanding frontiers of inquiry and application related to clinical research in depression and schizophrenia. It seems clear, nonetheless, that the development of a unifying conceptual framework for encompassing the fundamental features of these prototypically interactive phenomena must appeal to the scientific data base that defines the relationship between biological and behavioral mechanisms of action with precision and operational objectivity.

In this regard, it seems worth noting, for example, that there is more than a reasonable congruence between perhaps the most prominent of the proposed biological mechanisms for schizophrenia, the dopamine hypothesis (Synder 1978), and the more current behavioral formulations summarized previously. The "behavioral immediacy" generalization suggested by the constellation of schizophrenic performance deviations (i.e., behavior controlled by stimuli in the schizophrenic patient's immediate environment) is certainly consistent with presumed effects of dopamine excesses upon neural transmission along pathways and/or at times other than those normally activated by precipitating environmental events. Perhaps the action of the neuroleptic drugs in reducing the symptomatology associated with the suggested behavioral immediacy mechanism, while concurrently increasing dopamine-receptor blocking effects, provides some further support for this biobehavioral mechanistic congruence.

The clinical depressive disorders, complex in presentation, dissimilar in origins and course, and often pleomorphic in character, have long inspired theoretical discussion of behavioral mechanisms focused predominantly on conditioning and learning principles. Among the more popular accounts is the Seligman "learned helplessness" proposal (1975). Depressed individuals are presumed to have learned that nothing works (i.e., avoids punishment or produces reinforcement) in some fundamental situation and then to have generalized this to other situations that may otherwise have been manageable. In keeping with contemporary fashion, this rather basic learning mechanism hypothesis has lately been "cognitively" elaborated (or perhaps overelaborated), which in more operational terms is to say that verbal response-produced stimuli, covert as well as overt, are now assigned important causal roles. An alternative behavioral view would call attention to a more general decrease in the supportive role of reinforcing environmental consequences with a reduction in base levels of reinforcement affecting various areas of functioning (Lewinsohn 1975). Under such circumstances, the depressed behavior itself may produce environmental consequences that increase depression and eventually induce others to avoid the depressed individual, thus further constraining options for finding alternative sources of strength for more desirable behaviors.

It is self-evident that an adequate understanding of the depressive disorders, their origins, biological substrates, and amenability to established and novel forms of treatment can be advanced most effectively by experimental interventions that cannot readily or ethically be carried out in clinical populations. Perhaps one of the most useful complementary approaches to clinical research in depression has proven to be the preclinical behavioral models for laboratory investigation on which the conceptual formulations of Seligman and others have been based. Perhaps the most convincing arguments for the validity of such animal laboratory models are the clinically parallel changes in prominent behavioral and neuroendocrine abnormalities and their responsiveness to pharmacotherapy with standard tricyclic antidepressants as well as several structurally and pharmacologically atypical but clinically effective antidepressants. Moreover, the time course over which these effects are observed shows a high degree of correspondence with the clinical course of such interventions. The depressed behavior in such laboratory models is reversed only after chronic treatment with tricyclic antidepressants, atypical antidepressants, monoamine oxidase inhibitors, and electroconvulsive therapy. No effects are seen acutely with those agents. Anxiolytics, neuroleptics, barbiturates, alcohol, and central nervous system stimulants do not affect such behavioral depression models, either acutely or chronically. However, despite the strength of the argument for such laboratory behavioral models as useful and necessary adjuncts for developing an adequate understanding of depression in humans, their utility remains a direct function of a continuing dialogue between clinical and laboratory research. As such, these models demand scrupulous observation, experimentation, and methodological rigor on the part of both clinical and laboratory investigators.

Behavioral Neuroscience

To some extent, consideration of these behavioral mechanisms as characterizing depressive and schizophrenic disorders has lately been overshadowed by dramatic discoveries in microbiology and the neurosciences, particularly neurochemistry and neuropharmacology. Indeed, the relevance of explosive advances in new knowledge of neurotransmitter and receptor dynamics, as well as the developing enlightenment with regard to genetic predispositions, has been well documented in this book and elsewhere. Expansion of these knowledge bases can and does proceed independently of behavior analysis developments to the extent that appropriate methods and techniques provide for the observation and measurement of specifiable structural and/or functional entities. But before the relationship of these physicochemical events to the behavioral interactions defining psychopathological conditions can be usefully specified, both must be quantitatively described and shown to correspond in all their properties. This requirement becomes more cogent as effective independent methodologies and techniques are developed in the respective disciplines and, thus, must become more operational than conceptual (e.g., the waning popularity of the "conceptual nervous system" as we learn more about the real nervous system, in all but the diehard adherents of currently fashionable "cognitive science").

The point, of course, is that no amount of information about one domain will explain the other or bring order into it without the direct treatment represented by its own operational analysis. This argument applies equally well to all the medical disciplines related to behavior. No amount of exclusively endocrinological information, for example, will prove the thesis that "personality" is a matter of glandular secretion or that "thought" is neurochemical. What is required, if defense of such theses is to go beyond mere rhetoric, is a rigorous and independent behavioral formulation of what is meant by "personality" and "thought," including quantitative measurement of their properties. Only then can a legitimate and valid relational analysis between the two domains, inside and outside the skin, be productively pursued.

The relevance of these assertions should be immediately obvious. Not only must a science of behavior be independent of the neurosciences, but, in principle at least, it must be established as a separate discipline whether or not a rapprochement with neuroscience is productively pursued. Now this is not to gainsay the obvious advantages to simultaneous exploration of demonstrably related fields of inquiry. Rather, it is to make the point by emphasizing this seemingly isolationist view that the tendency to look inside the organism for an "explanation" of behavior has little in the way of hard evidence to recommend it. It is probably no exaggeration to suggest that such dispositions arise in large part from clinical practices wherein explanation has a relatively simple meaning. The finding of some organic malfunction as a correlate of some behavioral deficit is doubtless an important step in understanding the condition of a patient. But such success depends largely on the negative nature of the data, the absence or derangement of a function being much more easily described than the function itself. The point is that despite commonly held views that neuroscience facts somehow illuminate behavior or offer a simpler account of behavioral facts, there is little or no evidence that any such neurological findings have ever told anyone anything new about behavior (Adolf Meyer's "Neurologizing Tautology"). And from the point of view of a descriptive science, that is the only criterion to be taken into account.

I am well aware that the conception of interactive but independent and purely descriptive sciences of neurology and behavior, for example, as the fundamental building blocks of applied psychiatric medicine is not likely to find immediate favor with the rank and file of either researchers or practitioners. For those whose curiosity about nature is not equal to interest in the accuracy of their guesses, the hypothesis is the very lifeblood of science. Indeed, the opposition to pure description appears to be nowhere else as strong as in the disciplines concerned with clinical research and psychiatric medicine. But there are some redeeming features to this somewhat iconoclastic posture. First, independent definition of terms in the respective disciplines offers the tremendous advantage of keeping both investigators and clinicians aware of what is known and what is not known. Second, maintaining the independence of a scientific discipline keeps it free of unnecessary restraining influences, because it is clearly a snare and a delusion to burden one discipline with the difficulties inherent in another.

I hasten to provide assurances that I am not overlooking the advance that is made in unification of knowledge when terms at one level of analysis are defined or "explained," if you will, at a lower level. Eventually, a synthesis of the laws of behavior and of neuroscience may be achieved, but there are already clear signs that the reduction to lower terms will not stop at that level. We no longer have an anatomy department at the Johns Hopkins University School of Medicine, for example. It has been folded into a group identified as cell biology. And even as we speak, physiology departments all over the country are struggling for their existence in the face of patent reductionist influences.

Clearly, the final description will be in terms of whatever quasi-ultimate physical units are in fashion at the time. Under any circumstances, the intensive cultivation and rigorous prosecution of a field at its own level is to be strongly recommended, not only for its own sake but for the sake of more rapid progress toward integration as well as ultimate multidisciplinary synthesis. It must be recognized, of course, that the facts of behavior (or pseudofacts, as the case may be) appealed to by contemporary psychiatric medicine can be satisfying only in the earliest and most primitive—perhaps prescientific-developmental—stages of this basic and applied discipline.

Perhaps a more temperate view of the relation between the experimental analysis of behavior and the neurological sciences that I have been trying to convey can be extrapolated from the following quotation at the beginning of a chapter on physics and biology in Mach's *The Analysis of Sensations* (1914):

> It often happens that the development of two different fields of science goes on side by side for long periods, without either of them exercising an influence on the other. On occasion, again, they may come into closer contact, when it is noticed that unexpected light is thrown on the doctrines of the one by the doctrines of the other. In that case a natural tendency may even be manifested to allow the first field to be completely absorbed in the second. But the period of buoyant hope, the period of overestimation of this relation which is supposed to explain everything, is most often followed by a period of disillusionment, when the two fields in question are once more separated, and each pursues its own aims, putting its own special questions and applying its own methods. But on both of them the temporary contact leaves abiding traces behind. Apart from the positive addition to knowledge, which is not to be despised, the temporary relation between them brings about a transformation of our conceptions, clarifying them and permitting of their application over a wider field than that for which they were originally formed.

Conclusion

Notwithstanding what many regard as a less than praiseworthy commentary on the state of the art and science of clinical research and psychiatric medicine, it is abundantly clear that this endeavor—sufficiently new to generate both excitement and controversy, and more advanced in some areas than in others—is nonetheless responsible for important advances in applied biological science. The claims and disclaimers that continue to be heard in lecture halls and corridors will serve both to arouse and to guide our interest, but the major utility of our efforts must continue to stem from the background data and methodological descriptions they provide. We must remain committed to expanding the knowledge base necessary to provide direction and focus for the development of effective psychiatric medicine interventions. Research is the single most potent tool we have for accomplishing this expansion. Building a first-rate scientific image, seeking bridges, and fostering crossovers between basic and applied disciplines will elevate the entire psychiatric medicine enterprise. This fundamental approach will permit practitioners and researchers, both insiders and outsiders, to evaluate the current empirical status of the field and to judge how much is to be gained from continued efforts to pursue such interdisciplinary investments.

In this regard, a more rigorously operational behavioral paradigm would greatly enhance the knowledge base on which the success of such initiatives can be expected to depend. The caution that needs emphasis in this final plea, however, concerns the ongoing powerful influence of the mental health industry, supported by third-party payments, and the strong disposition to misdirect the focus of the field inward—creeping cognitivism, in league with the traditionally classic approach of blaming the victim. This would not be good news for the consumer-patient and certainly not for the operationally oriented research clinician. We must, at all costs, guard against the fractionation of psychiatric medicine and empirical behavioral science. The resulting disappearance of the behavioral scientist from the scientist-practitioner model would sound the death knell of the field as we know it.

References

Blakely E, Schlinger H: Rules: function-altering contingency-specifying stimuli. Behavior Analysis 10:183–187, 1987

Estes NK: Report of a research briefing on cognitive science and artificial intelligence, in Research Briefings 1983. Washington, DC, National Academy Press, 1983

Frank JD: Persuasion and Healing. Baltimore, MD, Johns Hopkins University Press, 1961

Lewinsohn PM: The behavioral study and treatment of depression, in Progress in Behavior Modification, Vol 1. Edited by Hersen M, Eisler RM, Miller PM. New York, Academic, 1975

Mach E: The Analysis of Sensations. Chicago, IL, Open Court Publishing, 1914

Pomerleau OF, Brady JP: Behavioral Medicine: Theory and Practice. Baltimore, MD, Williams & Wilkins, 1979

Salzinger K: Schizophrenia: Behavioral Aspects. New York, Wiley, 1973

Salzinger K: The behavioral mechanism to explain abnormal behavior. Ann N Y Acad Sci 340:66–87, 1980

Salzinger K: The immediacy hypothesis in a theory of schizophrenia, in Nebraska Symposium on Motivation: Theories of Schizophrenia and Psychosis. Edited by Spaulding WD, Cole JK. Lincoln, NE, University of Nebraska, 1984

Salzinger K, Portnoy S, Feldman RS: Verbal behavior and some comments toward a theory of schizophrenia, in Psychopathology of Schizophrenia. Edited by Hock P, Zubin J. New York, Grune & Stratton, 1966

Salzinger K, Portnoy S, Pisoni, DB, et al: The immediacy hypothesis and response-produced stimuli in schizophrenic speech. J Abnorm Psychol 76:258–264, 1970

Salzinger K, Portnoy S, Feldman RS: Communicability deficit in schizophrenics resulting from a more general deficit, in Language and Cognition in Schizophrenia. Edited by Schwartz S. Hillside, NJ, Erlbaum Associates, 1978

Seligman MEP: Helplessness: On Depression, Development and Death. San Francisco, CA, Freeman, 1975

Simon H: The behavioral and social sciences. Science 209:72–76, 1980

Snyder SH: Dopamine and schizophrenia, in The Nature of Schizophrenia: New Approaches to Research and Treatment. New York, Wiley, 1978

7

Molecular Genetics and Psychiatry

Floyd E. Bloom, M.D.

The results of recent efforts in psychiatric epidemiology strongly support the notion that genetically transmittable factors underlie the enhanced susceptibility to affective disorder, schizophrenia, and alcoholism in certain individuals (see Pardes et al. 1989). Such data could be taken to imply a new era of biological psychiatry, in which the molecular explanations of a biologically linked psychosis might be fruitfully understood through examination of the molecular genetic specifications of the brain. However, two general precautions should be recognized before exulting in this soon-to-be-entered era of specific molecular mechanisms of psychosis:

1. The exact nature by which neurons are linked chemically into systems that provide the means for complex human cognitive and emotional operations are unknown. Thus, we are unlikely at any time soon to understand how such operations deteriorate in psychotic patients, even with more molecules at hand to study.
2. We can already be sure from the recent partial progress with the biochemical analysis of the Lesch-Nyhan defect (see Martin 1987 for additional references) that even the precise identification of a genetic defect specifically linked to a behavioral/cognitive disease that is expressed through the dysfunction of an identified protein will not by itself "explain" how the enzyme and the neurons that express it relate to the behavioral problem. In part, these problems are immense but, more simply, we may merely acknowledge that our understanding of the brain and the molecules and processes by which it functions remain at an improved but still primitive state.

Supported by NIH Grant NS 22347.

One approach to defining the critical elements of such a molecular genetic specification emphasizes the detection and characterization of the proteins underlying the phenotypic properties of brain cells and the emergent functions attributable to their multicellular ensembles. However, most of the complex features of the human brain's organization remain incompletely defined. Furthermore, the essential properties of this brain that enable it to achieve the cognitive, emotional, and other behavioral performance characteristics that signify the symptoms of psychosis are unknown (see Pardes et al. 1989), other than the increase in cortical mass and the added amount of cortico-cortical interconnectivity that is thereby provided.

Therefore, to begin to develop a research strategy that could potentially illuminate these unknown proteins, functions, and supracellular attributes, we have combined molecular biological, biochemical, and cytological methods to survey the degree to which elements of the mammalian genome are selectively expressed in the brain. Our early results (Milner and Sutcliffe 1983; Sutcliffe et al. 1983) suggest that at least half of the genome may be viewed as "brain-specific" and that the number of possible proteins pertinent to brain function are too numerous to examine by routine selection. From that task-defining realization has emerged a still more refined strategy to analyze the molecular basis of functional properties within defined fields of the primate neocortex and during defined stages of brain development. We view the identification of these molecules as a critical step in approaching mechanisms pertinent to characterization of the human brain and its neuropsychiatric dysfunctions. Functional and morphologic properties have been the traditional indices by which neurons and their circuits are classified (Jones and Hendry 1988). Although such descriptive schemes document neuronal diversity, they cannot per se account for the variations in properties or the nature of the molecular signals that guide and establish these connections and their functions.

Efforts to clarify the uniquely complex structural organization of the brain were first based on empirically derived cell staining methods supplemented later by more powerful and specific circuit tracing methodologies. During the past decade, these anatomic studies have been powerfully extended by immunocytochemical probes specific for products known or inferred to be produced by neurons or glia. The majority of known neurotransmitter substances and their related metabolic enzymes and receptors are shared among the nervous, endocrine, and immune systems. Although such messenger molecule markers can offer extremely pertinent insights into genetically transmitted diseases of the brain (see Martin 1987, 1989), the study of transmitters alone cannot provide a comprehensive approach to the determination of the molecular basis of cellular specificity in the brain because of the high proportion of neurons whose transmitters have not yet been identified.

Attempts to define the specific biochemical properties of the nervous system have generally begun with comparative analysis of its chemical differences with cells of other organ systems, especially with regard to the unique lipids of the brain (Lees and Brostoff 1984) and the transmitter gamma-amino butyrate (see Roberts and Kuriyama 1968). A direct biochemical approach has also been highly successful in identifying specific molecules based on existing assays of predicted action, such as the hypophysiotrophic hormones (see Guillemin 1978) or based on certain predictable chemical features of transmitter peptides, such as C-terminal amidation (see Tatemoto and Mutt 1980). However, by their specialized nature, it is difficult to infer from these successful applications of an assay-based identification scheme whether or not the specific molecules identified represented a large or small proportion of the brain's available physiological regulatory systems. Although the lists of brain-specific, glia-specific, or neuron-specific molecules that have been detected is growing (see Sutcliffe 1988), we are still in the primitive stages of such list-building, let alone of recognizing its functional implications.

Molecular Biological Studies on the Rat Brain

We have, therefore, sought a more direct route to the molecular basis by which such neural properties are generated and maintained. Such an alternative approach has only recently become accessible through the powerful methods of recombinant DNA technology. As noted by Sutcliffe (1988), the past 5 years have witnessed an advantageous coupling of the methods of molecular biology to neuroscience. This combination has provided a general solution to the initially overwhelming complexity of the brain by making available rapid and precise means to reduce the enormous numbers of molecules to the many or few that may meet an operational definition of what a specific investigative protocol may wish to define as "interesting." By providing power, speed, precision, and the ability to define the target molecules of interest, mammalian neuroscientists can in principle create their preferred model systems of molecular discovery without relinquishing their desire to obtain data on the mammalian brain itself.

Using these methods, our original efforts were directed toward the general issues of whether the special properties of the brain were based on different proportions of the same proteins and other gene products found in other organs or whether the special functional properties of the brain were derived from genes and gene products that might be regarded as specific for brain function (Milner and Sutcliffe 1983; also, see Milner et al. 1987). These brain messenger RNA (mRNA) complexity studies extended earlier observations (see Chaudhari and Hahn 1983; Kaplan and Finch 1982) that the mammalian brain expresses a large number of distinct mRNAs, expressed exclusively or at least predominantly in the brain as a whole and, in most regions, increasing in abundance in the brains of those species whose brains develop postnatally (see Milner et al. 1987 for more extensive references).

The majority of the clones derived from mRNAs expressed in the adult rat brain as reported by Milner and Sutcliffe (1983) could be reduced to a few distinct categories, based on the degree to which these mRNAs were also detectable in the adult liver or kidney. Those defined as Class I mRNAs, accounting for about 20% of the total, were detectable in approximately identical abundances in brain, liver, and kidney. This category of mRNA may be conceived of as encoding "housekeeping proteins," inferred to be expressed in all cells at approximately the same level. One such example, the cyclosporin-binding protein, "cyclophilin" (Danielson et al. 1988), was detected from this strategy and has served as a useful internal normalization index of mRNA preparations from a variety of organs and brain regions. A second large class of brain-derived mRNAs reported by Milner and Sutcliffe (1983), or Class II, also represented about 20% of the total in brain and were also present in the liver or kidney, but at abundance estimates that differed significantly across these organs. This class is inferred to represent those mRNAs that encode proteins needed in more than one cell type but at different concentrations, depending on the degree of specialization of the cell and the protein. An example of such mRNAs are those that encode the structural proteins of the tubulin family (also see Miller et al. 1987b).

The third class of brain-derived mRNAs defined by Milner and Sutcliffe (1983), Class III, were those detected in brain but not liver or kidney. Thus, operationally, these mRNAs and their protein products can be regarded as brain specific. In comparisons among the three categories of brain-derived mRNAs, the Northern blot analyses revealed that those of Class III were longer by almost twofold than those of Class I; those Class III mRNAs of lesser abundance tended to be the longest of this class. Given that the rarest detectable Class III mRNAs averaged about 5,000 nucleotides in length and made up the bulk of those derived from the brain numerically, Milner and Sutcliffe (1983) have estimated that the brain must express at least 30,000 genes, of which at least 30% must be regarded as enriched in or selective for the brain.

Using these methods, the specific but unknown properties of neurons, both generally and individually, can be approached directly in terms of the gene products expressed in brain but not in other large minimally innervated organs (see Sutcliffe et al. 1983, 1984). With this basic organ differential approach, we sought to develop a more general approach to analysis of the molecular basis of brain specificity. We isolated complementary DNA (cDNA) clones of several mRNAs that are enriched in brain by more than 100 times the limits of detectability for liver and kidney and determined their nucleotide sequences, thereby obtaining the amino acid sequence of the corresponding proteins by deduction from the genetic code.

The presumptive gene products are then detected using polyclonal antisera raised against selected synthetic fragments of the deduced amino acid sequences (Sutcliffe et al. 1983). In addition to revealing immunocytochemically defined shared properties of otherwise unrelated neurons in widely separated portions of the rat neuraxis, antisera to the peptide products of these brain-specific mRNAs have also provided useful markers of brain differentiation (Lenoir et al. 1986; Miller et al. 1987a, 1987b) and have allowed for ready comparison across species.

One specific clone analyzed in this fashion was the rat brain-specific protein 1B236. This clone was originally defined by characterization of randomly selected cDNA clones of mRNAs expressed in adult rat brain but not detectable in liver or kidney (Milner and Sutcliffe 1983; Sutcliffe et al. 1983).

Nucleotide sequence analysis of p1B236 provided the 3′ partial sequence of the mRNA, revealing the 3′ end of an open reading frame from which a 318 amino acid putative translation product was deducted. Antisera to each of three synthetic peptides corresponding to nonoverlapping regions of the C-terminus of the 1B236 sequence detected a 100 kd rat brain protein containing up to 30 N-linked carbohydrate sites (also see Malfroy et al. 1985). Proteolytic fragments derived from the C-terminus, corresponding to three of the synthetic peptides predicted as possible cleavage products, were also detectable (Malfroy et al. 1985). Both 1B236 mRNA and its protein are expressed early in rat brain development, beginning on postnatal day 3 (see Lenoir et al. 1986). Immunocytochemical analysis (Bloom et al. 1985) and in situ hybridization analysis (Higgins et al. 1989; Lai et al. 1987a) showed two phases of development expression. During days 3 through 20, 1B236 is expressed predominantly in oligodendrocytes within myelinating tracts throughout the central nervous system (CNS). However, after day 20, 1B236 mRNA and protein are detected predominantly in subsets of neurons within gray matter of the olfactory, limbic, motor, and somatosensory systems (Bloom et al. 1985; Higgins et al. 1989; Lai et al. 1987a).

Further analysis revealed that there is only one 1B236 gene, but at least two major forms of mRNA, that can be derived from it by differential splicing (Lai et al. 1987a). The alternative form also encodes two slightly different forms of the protein product, which vary at their C-termini. From the complete primary structure of the two major RNA forms, we deduce that both protein products have a single transmembrane region, separating a large, highly glycosylated amino terminal region from two alternative carboxyl terminal tails. The shared amino terminal region consists of five domains of roughly equal size, each encoded by a separate exon, that are closely related in sequence to each other. These domains of internal similarity also reveal further homologies by computer search of previously known protein structures. In particular, 1B236 was found to show significant structural similarity to members of the immunoglobulin superfamily (Williams 1987), including the neural cell adhesion molecule, N-CAM (Cunningham et al. 1987), and the platelet derived growth factor (Yarden et al. 1986; also see Lai et al. 1987b). In addition, through cross-reactivity of in vitro expressed segments of the 1B236 mRNAs, the 1B236 protein is indistinguishable from the so-called myelin associated glycoprotein (MAG), a chemically defined, nervous system specific glycoprotein expressed by oligodendrocytes during the process of myelination (Quarles 1984) and purported to be functionally related to the interaction between the myelin producing cells and the axons they myelinate (Martini and Schachner 1986).

From the work to date on 1B236/MAG, one recognizes that many brain-specific proteins exist for which no previously conceived functions could have predicted the sorts of bioassays that have classically led to molecular discovery. Given that 1B236 was selected for analysis virtually at random, the likelihood of encountering interesting molecules by chance might seem to be reasonably high. Thousands of additional mRNAs have emerged from the same series of brain mRNA detection experiments (Sutcliffe et al. 1983); more than 2 dozen are presently under active examination. Nevertheless, it became clear to us that additional strategies were required to focus our attack.

There is no lack of strategies by which one might begin to analyze the pool of brain-specific mRNAs for those that encode protein products that would assist in the orderly dissection of the cells and cell systems in the brain. For example, Greengard and collaborators have developed the strategy of detecting proteins that serve as substrates of phosphorylation or dephosphorylation reactions; they have used this approach to characterize several previously undisclosed proteins within definable chemical phenotypes (Walaas et al. 1983a, 1983b). We opted for a strategy that relied more on brain cellular structure, because the past decade's research on neuronal geometry has made clear that even chemically similar neuronal phenotypes can express wide degrees of structural variation, which presumably reflects some uncertain aspect of innervation density (Purves and Lichtman 1985).

Therefore, we have evolved a strategy from the premise that subsets of neurons defined by their grouping of shared or unique functional features could help define the molecular markers of these functions. By trying to identify the molecular systems underlying the characteristic structural and functional properties of the neurons of the primate cerebral cortex, we sought a more restrictive but still functionally open-ended approach to identifying cell-specific functional molecules. Certainly, the neocortex is a critical region for understanding the sorts of dysfunctions unique to the human brain, such as the psychoses on which this volume focuses attention. Furthermore, several obvious features of the neocortex were directly relevant for our focused objectives. From our operationally oriented perspective, we viewed the cerebral cortex as a brain region in which vastly differing functional capacities are expressed by neurons arranged in macroscopically homogeneous arrays that differ only modestly from one region of cortex to another, but in which there are important variations in function, in cytoarchitectonics, and in connectivity properties within functionally definable regions. The added complexity of the primate cerebral cortex, defining even greater regional and subregional specializations, with each having its own more highly refined cytoarchitectonic basis, suggested that this cortex was a highly suitable system in which to exploit molecular biologic approaches to neural cell specificity as well as of obvious importance for understanding overall brain function and cognitive dysfunction.

Using conventional methods, poly A-positive mRNA was isolated from young adult male cynomolgus monkeys (see Travis et al. 1987). The initial primate brain cDNA library generated from this RNA yielded more than 1 million clones, each with an insert frequency of approximately 90% and an insert size ranging from several hundred to at least 6,000 mRNAs. Given this imposing number of primate cortex-derived mRNAs, a two-step strategy was devised for selecting those of functional importance for neocortex as compared to cerebellum: 1) employing subtractive hybridization methods (cortex derived 32P labeled single stranded cDNAs were hybridized to a large excess of unlabeled cerebellar derived mRNA from the same specimens), and 2) employing differential colony hybridization, with cDNA prepared from either cortex or cerebellum to define those from the first stage studies that were expressed selectively in neocortex (see Travis et al. 1987). One typical run, with approximately 25,000 cDNA clones derived from mRNAs of the adult primate neocortex, first yielded approximately 1,200 that were apparently enriched in neocortex over cerebellum and reduced to fewer than 50 cDNAs when these "cortex enriched" clones were further defined with greater sensitivity after being hybridized with probes of cDNAs derived from cerebellum or liver. The resultant set of tentative cortex-specific mRNAs was then screened for intracortical regional abundance variation by performing Northern blots with these clones on mRNAs derived from either visual, motor, or dorsal prefrontal cortex.

Of the first several mRNAs of this last stage, five were found by these criteria to be selective for cortex and not detected in cerebellum, one detected mRNAs of different sizes in neocortex and cerebellum, and three others were found in mRNAs of several other brain regions but were relatively reduced in cerebellum. One of those initially regarded as present in cortex and not in cerebellum (termed 1B4 in our protocols) was, in fact, also found in cerebellum following direct Northern blot analysis with cerebellum derived cDNA, was at slightly greater overall abundance in motor or dorsal prefrontal cortex as compared to visual cortex, and was later found to be markedly reduced in basal ganglia (see Travis et al. 1987).

Of the initial, independently detected cortex "specific" clones (compared to cerebellum or liver), all five were later found to hybridize to target mRNAs of approximately the same size and with an abundance of approximately 0.05%; and all five showed cross-hybridization on Southern blot analysis, indicating that they represented the same brain-specific, cortex- specific gene—1H8A—in our protocols (see Travis et al. 1987). To define in greater spatial precision the cellular sites in which these cortex enriched mRNAs were expressed, we applied in situ hybridization using 35S plasmid derived DNA. Clone 1H8A detected mRNAs in subpopulations of cells, in laminae II–VI, in all cortical regions examined, with slightly higher densities in prefrontal regions over temporal regions. Clone 1B4, although not meeting our desire for cortical exclusivity, showed far greater selective spatial distribution across cortical regions and laminae on in situ hybridization analysis. Labeling of cells was most dense in laminae V and VI of the primary visual region (area 17) of the occipital lobe and in the inferior temporal gyrus of the temporal lobe, as well as in layered neurons of the underlying lateral geniculate nucleus (see Travis et al. 1987).

More recently, using still further refinements of this two-step subtractive-differential hybridization approach (Travis and Sutcliffe 1988), the overall sensitivity of the method has been further increased to extend the threshold for detection down to at least 0.001% of the total mRNA. Following this methodological improvement, it has again been possible to identify cortex derived mRNAs not apparently expressed in cerebellum. An initial breakdown of 100 clones meeting this operational, differential region detection criterion (G.H. Travis, M.D., and J.G. Sutcliffe, Ph.D., personal communication, 1988) indicates that approximately 93 isolates were again representing the mRNA of 1H8A (with an overall abundance of 0.05% or more). Of the other seven primate cortex derived isolates, whose abundance approaches the current theoretical threshold of 0.001%, six have undergone initial nucleotide sequence analysis, revealing that two represented the neuropeptides somatostatin and cholecystokinin, known to be absent from cerebellum, as well as four other unique sequences of approximately the same abundance. The latter show some regional variations within cortex on Northern blot analyses, but none so far analyzed can be regarded as unique to any primate neocortical region. These cDNAs remain under active investigation as to their cellular distributions. In all cases, detailed regional as well as cellular mapping, both with in situ hybridization as well as by immunocytochemistry with antisera raised against the gene product, will be necessary to draw any definitive conclusions regarding breadth or selectivity of their distribution.

Based on the data we currently have, a very conservative interpretation of the results would take the view that, thus far, no "cell-type" specific mRNAs have been detected by our approaches. If this conclusion is supported by further observations, it would suggest that the obvious and dramatic differences in neuronal cytology, circuitry, and signaling systems are based on recombinations in varying proportions and with varying representations of members of the rare to very rare classes of brain mRNAs, with many proteins (even those defined as neuron specific) being represented more frequently and most broadly. Given this array of common, less common, and rare mRNAs, some of which have begun to be defined by our methods as region-specific, one might still speculate that such region-specific gene products could provide important determinants on the neurons within the region.

Conclusions

At this stage in the evolution of the neurosciences, molecular and cellular research in combination provide a progressive, interactive series of research strategies potentially capable of defining specific neuronal phenotypic markers and the developmental patterns by which they are expressed uniquely, as well as the possible principles by which they may be shared. The strategies we have undertaken to develop have certain operational advantages in conceptualization of disorders (such as the psychoses) in which the nature of the underlying factors enhancing susceptibility to express the signs and symptoms of the disorder must be based on CNS abnormalities. Furthermore, it is our contention that identification of brain-specific gene products without bioassay filters for the discovery process can, in fact, reveal new cellular functions and novel molecularly based concepts of brain organization.

Another example from our studies is one that may perhaps be more closely related to the type of very subtle changes that could be viewed as pertinent to models of psychosis-gene transmission. In studies of gene expression in the postnatally developing rat hypothalamus, using the mRNA for the oxytocin gene as a probe, we have observed that there is a marked gender-based and hormone sensitive regulation of this neuropeptide, such that the male brains are far richer and have many more neurons expressing the oxytocin mRNA, whereas in the female, puberty regulates the amount of oxytocin expression in a restricted number of neurons (Miller et al. 1989). Because at least one of the forms of chronic schizophrenia has been classically linked to brain changes beginning at or near puberty, it is reasonable to consider that far greater definition of these effects of gonadal maturation on the brain must be pursued.

A comprehensive strategy to elucidate the molecular basis of neuronal specificity thus constitutes an important long-term goal of our research program. As a part of the broadly based strategies currently evolving to link genetic markers with complex human disorders, we take the view that any of the thousands of still-to-be-discovered brain-specific genes could provide a genomic mapping marker of direct CNS pertinence, and that this same application could be made with more emphasis for those that can be further characterized as region-specific. Given the very large number of gene defects that can result in "mental retardation" states (see Gurling 1986), it may be presumed that normal cognitive ability must depend on a very large array of gene products. Because the incidence of the major psychoses can be linked only partially to inheritable factors, we intend to persist in our general strategy. Although we are sanguine that cell specific markers may not yet be definable by these methods, should they exist, it remains likely that knowledge of brain-specific and region-specific mRNAs will nevertheless offer important insight into the organizational and functional complexity of the brain.

References

Bloom FE, Battenberg ELF, Milner RJ, et al: Immunocytochemical mapping of 1B236, a brain specific neuronal polypeptide deduced from the sequence of a cloned mRNA. J Neurosci 5:1781–1802, 1985

Chaudhari N, Hahn WE: Genetic expression in the developing brain. Science 220:924–928, 1983

Cunningham BA, Hemperly JJ, Murray BA, et al: Neural cell adhesion molecule: Structure, immunoglobulin-like domains, cell surface modulation, and alternative RNA splicing. Science 236:799–806, 1987

Danielson PE, Forss-Petter S, Brown MA, et al: p1B15: a cDNA clone of the rat mRNA encoding cyclophilin. DNA 7:261–267, 1988

Guillemin R: Peptides in the brain: the new endocrinology of the neuron. Science 202:390–398, 1978

Gurling H: Candidate genes and favoured loci: strategies for molecular genetic research into schizophrenia, manic depression, autism, alcoholism, and Alzheimer's disease. Psychiatr Dev 4:289–309, 1986

Higgins GA, Schmale H, Bloom FE, et al: Cellular localization of 1B236/myelin-associated glycoprotein (1B236/MAG) mRNA during rat brain development. Proc Natl Acad Sci U S A 86:2074–2078, 1989

Jones EG (ed): Expression of neuronal diversity in the central nervous system, in Molecular Biology of the Human Brain (UCLA Symposia of Molecular and Cellular Biology, Vol 72). Edited by Jones EG. New York, Alan R. Liss, Inc., 1988

Kaplan BB, Finch CE: The sequence complexity of brain ribonucleic acids, in Molecular Approaches to Neurobiology. Edited by Brown IR. New York, Academic, 1982

Lai C, Brow MA, Nave K-A, et al: Two forms of 1B236/MAG, a cell adhesion molecule for postnatal neuronal development are produced by alternative splicing of the 1B236/Myelin-associated glycoprotein (MAG) gene. Proc Natl Acad Sci U S A 84:4337–4341, 1987a

Lai C, Watson J, Bloom FE, et al: The neural protein 1B236/MAG defines a subpopulation of the immunoglobulin super family. Immunol Rev 100:129–151, 1987b

Lees MB, Brostoff SW: Proteins of Myelin, in Myelin. Edited by Morell P. New York, Plenum, 1984

Lenoir D, Battenberg E, Kiel M, et al: The brain-specific gene 12B36 is expressed postnatally in the developing rat brain. J Neurosci 6:522–530, 1986

Malfroy B, Bakhit C, Bloom FE, et al: Brain-specific polypeptide 1B236 exists in multiple molecular forms. Proc Natl Acad Sci U S A 82:2009–2013, 1985

Martin JB: Molecular genetics: applications to the clinical neurosciences. Science 238:765–772, 1987

Martin JB: Molecular genetics studies in the neuropsychiatric disorders. Trends Neurosci 12:130–138, 1989

Martini R, Schachner A: Immunoelectron microscopic localization of neural cell adhesion molecules (L1, N-CAM, and MAG) and their shared carbohydrate epitope and myelin basic protein in developing sciatic nerve. J Cell Biol 103:2439–2448, 1986

Miller FD, Naus CCG, Higgins GA, et al: Developmentally regulated rat brain mRNAs: molecular and anatomical characterization. J Neurosci 7:2433–2444, 1987a

Miller FD, Naus CCG, Higgins GA, Bloom FE, Milner RJ: Isotypes of alpha-tubulin are differently regulated during neuronal maturation. J Cell Biol 105:3065–3073, 1987b

Miller FD, Ozimek G, Milner RJ, et al: Regulation of neuronal oxytocin mRNA by ovarian steroids in the mature and developing hypothalamus. Proc Natl Acad Sci U S A 86:2468–2472, 1989

Milner RJ, Sutcliffe JG: Gene expression in rat brain. Nucleic Acids Res 11:5497–5520, 1983

Milner RJ, Bloom FE, Sutcliffe JG: Brain specific genes: strategies and issues, in Current Topics in Developmental Biology, Vol 21. London, Academic, 1987

Pardes H, Kaufmann CA, Pincus HA, et al: Genetics and psychiatry: past discoveries, current dilemmas, and future directions. Am J Psychiatry 146:435–443, 1989

Purves D, Lichtman J: Geometrical differences among homologous neurons in mammals. Science 228:298–305, 1985

Quarles RH: Myelin-associated glycoprotein in development and disease. Dev Neurosci 6:285–303, 1984

Roberts E, Kuriyama K: Biochemical-physiological correlations in studies of the gamma-aminobutyric acid system. Brain Res 8:1–37, 1968

Sutcliffe JG: mRNA in the brain mammalian central nervous system. Annu Rev Neurosci 11:157–198, 1988

Sutcliffe JG, Milner RJ, Shinnick TM, et al: Identifying the protein products of brain-specific genes with antibodies to chemically synthetized peptides. Cell 33:671–682, 1983

Sutcliffe JG, Milner RJ, Gottesfeld JM, et al: Control of neuronal gene expression. Science 225:1308–1315, 1984

Tatemoto K, Mutt V: Isolation of two novel candidate hormones using an ethical method for finding naturally occurring polypeptides. Nature 285:417–419, 1980

Travis GH, Naus CG, Morrison JH, et al: Subtractive cDNA cloning and analysis of primate neocortex mRNAs with regionally-heterogeneous distributions. Neuropharmacology 26:845–854, 1987

Travis GH, Sutcliffe JG: Phenol emulsion-enhanced DNA-driven subtractive cDNA cloning: isolation of low-abundance monkey cortex-specific mRNAs. Proc Natl Acad Sci U S A 85:1696–1700, 1988

Walaas SI, Nairn AC, Greengard P: Regional distribution of calcium- and cyclic adenosine 3′:5′-monophosphate-regulated protein phosphorylation systems in mammalian brain, I: particulate systems. J Neurosci 3:291–301, 1983a

Walaas SI, Nairn AC, Greengard P: Regional distribution of calcium- and cyclic adenosine 3′:5′-monophosphate-regulated protein phosphorylation systems in mammalian brain, II: soluble systems. J Neurosci 3:302–313, 1983b

Williams AF: A year in the life of the immunoglobin super-family. Immunol Today 8:298–303, 1987

Yarden Y, Escobedo JA, Kuang W-J, et al: Structure of the receptor for platelet-derived growth factor helps define a family of closely related growth factor receptors. Nature 323:226–232, 1986

8

Antipsychotic Drugs as Tools for Etiological Research

Arvid Carlsson, M.D.

Although in this chapter I will start out by reviewing current knowledge and speculations about the role of dopamine in schizophrenia, I will attempt to widen that perspective. I will discuss the function of dopamine in a broader context, focusing on the interaction with other neurotransmitters. Schizophrenia will be looked upon here as a syndrome, resulting from an imbalance among a number of different neurotransmitters or neuronal systems in the brain.

The dopamine hypothesis of schizophrenia contains two different parts, one dealing with the pharmacology and the other with the chemical pathology of this disorder (for a review, see Carlsson 1988). Whereas the former rests on solid ground, the latter is still a matter of some controversy. In case the dopaminergic system should turn out not to be primarily affected in schizophrenia, which cannot yet be determined, we are facing a situation reminiscent of Parkinson's disease before the discovery of the dopaminergic deficiency in that condition. Anticholinergic agents were known to have a therapeutic, though limited, efficacy in Parkinson's disease; but there was no evidence of a primary disturbance in the cholinergic system. In schizophrenia, we are dealing with what is evidently an imbalance between the dopaminergic and other systems in favor of the former. In case this analogy turns out to be adequate, it would seem logical, in the case of schizophrenia, to search for an abnormality in a system operating in intimate association with the dopaminergic system.

This review will be confined to the pharmacological aspect of the dopamine hypothesis. It is not necessary to recapitulate all the evidence. However, I would like to emphasize the close parallelism between motor and mental changes induced by manipulations of the dopaminergic system (Nauta 1989), which, I believe, calls for a revival of the classical term "psychomotor activity." If dopamine stores are depleted or if dopamine receptors are blocked by drugs, both mental and motor functions are strongly inhibited. In experimental animals, the picture of catalepsy is produced; that is, the animal is essentially immobile and may remain in the most awkward positions in which it is placed. In human beings, when dopaminergic function is reduced, the flow of thoughts and associations is strongly reduced, and the mood may be depressed. In fact, these mental symptoms tend to show up at somewhat lower dose levels than the motor phenomena, which is fortunate in a therapeutic context, even though the safety margin in this respect is narrow. When dopaminergic functions are elevated by releasing agents such as amphetamines or by direct receptor agonists, psychomotor activity is elevated. At first this elevated activity occurs in an apparently purposeful manner and is accompanied by an elevated mood. At higher dose levels, however, both motor and mental functions appear to undergo a disintegration, resulting in so-called motor stereotypes as well as in a corresponding cognitive disintegration, which ultimately lead to psychotic states, some of which are reminiscent of schizophrenia.

It seems reasonable to assume that the mental changes induced by pharmacological manipulation of the dopamine system involve the cerebral cortex and, perhaps especially, the prefrontal cortex. As a result, a number of workers in this field have postulated that the antipsychotic action of neuroleptics, as well as the psychotogenic action of dopaminergic agonists, resides in the cerebral cortex. I believe that this is, at best, only partly true and that an alternative explanation exists for the mental functions of dopamine.

My main argument is simply that the level of dopamine in the human cortex is extremely low; and this is also true of the density of dopamine receptors, especially the D-2 receptors, on which the major antipsychotic agents predominantly act. These data are supported by postmortem measurements on the human brain as well as by some recent positron-emission tomography (PET) data from the Karolinska Institute (see Carlsson 1988; Farde et al. 1988). From studies comparing human beings with lower primates, data indicate that an enormous evolutionary jump has taken place in terms of cortical growth. The dopamine system of the cerebral cortex, however, has not kept pace, suggesting that the mesocortical dopamine system has lost importance.

Striatal Connections

Thus, we have good reasons to look for a mechanism by which subcortically located dopamine can influence cortical functions. When trying to do so, we must examine the neuroanatomy of the basal ganglia and its connections. Fortunately this research area has made considerable advances in recent years.

We propose that the cerebral cortex is capable of controlling its sensory input and arousal by means of a negative feedback loop involving the striatal complexes, the thalamus, and the mesencephalic reticular formation (see Figure 8–1). This feedback loop is assumed to comprise several paths, representing motor, cognitive, and emotional functions. Each of these probably can be subdivided into different functional parts. For example, Narabayashi (1988) has observed that, by means of very small electrolytic lesions (2–3 mm in diameter) applied in the thalamus of Parkinson patients, it is possible to selectively alleviate tremors by placing the lesions in the ventrointermedial nucleus, whereas rigidity can be eliminated by placing the lesions in the ventrolateral nucleus. Larger lesions in this area frequently lead to mental complications. In the future, it may be possible to obtain a more complete map of the pathways and nuclei involved in the various components of motor and mental functions within this complex feedback system, which would also take into account the functional asymmetry of the cerebral hemispheres.

In this context, it is appropriate for me to comment briefly on the functions of the dorsal versus ventral striatum. The statement has been made that

Figure 8–1. Schematic drawing illustrating the hypothesis that the cerebral cortex is capable of controlling its sensory input/arousal via a dopamine-modulated feedback loop involving the striatal complexes and the thalamus/mesencephalic reticular formation.

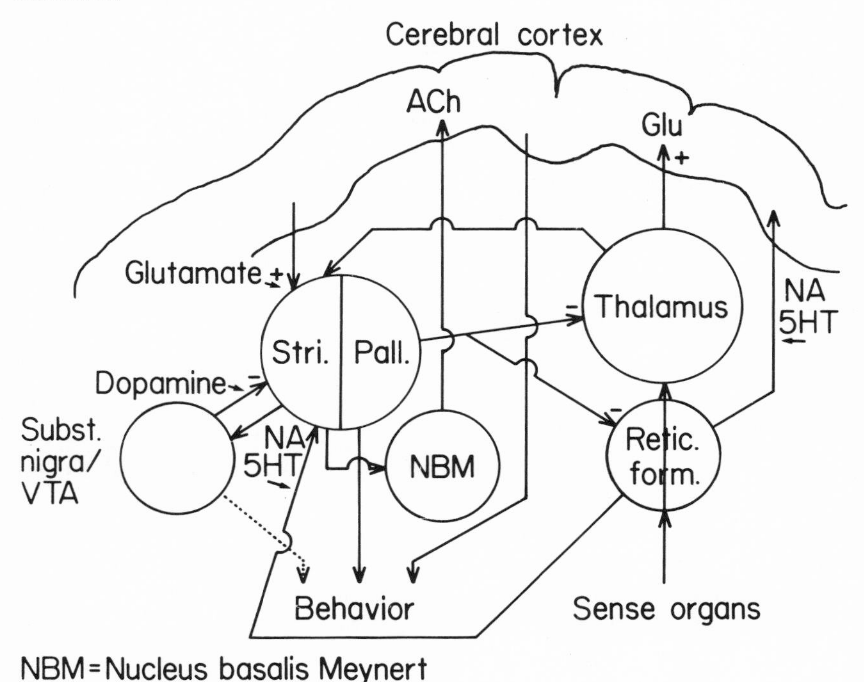

the former is a motor and the latter a "mental" structure. However, the anatomical connections of these structures suggest a more complex picture. For example, the dorsal striatum (essentially the caudate nucleus and the putamen) can, with respect to its connections, be divided into an anterior part, which is mainly connected to the association areas of the cortex, and posterior part, which, judging from its connections with the motor cortex, is predominantly a motor structure. The ventral striatum consists of a number of dopamine-rich structures located ventrally of the caudate nucleus and putamen. The nucleus accumbens is one important component of the ventral striatum. It should be emphasized that, in the human brain, the counterpart of this nucleus in lower mammals extends into the caudate nucleus and actually forms a major, ventromedial part of this nucleus (Alexander et al. 1986; Björklund and Lindvall 1986; Goldman-Rakic and Selemon 1986; Heimer et al. 1985; Nauta 1989; Selemon and Goldman-Rakic 1985).

The afferent and efferent connections of the dorsal and the ventral striatum are similar in principle, although different parts of the brain are involved. The afferents are largely derived from the cortex. However, the dorsal striatum receives its supply largely from the neocortex, whereas the ventral striatum is mainly innervated by the limbic cortex. In either case, we are apparently dealing with glutamatergic fibers. This glutamatergic system is a primary focus of this review. There also seems to be an important glutamatergic supply from the thalamus.

The efferents from the striatum project mainly to the pallidum, which can in turn be divided in a dorsal and a ventral part, communicating with the dorsal and the ventral striatum, respectively. Important further projections go to the thalamus. However, the dorsal and the ventral pallidum innervate different thalamic nuclei. Whereas the dorsal projections go to the ventrolateral nucleus, which appears to project to the neocortex in a precisely targeted manner, the ventral pallidum projects to nuclei in the thalamus belonging to a diffuse system capable of controlling vast areas, including the frontal and limbic cortex and the amygdala (Nauta 1989).

Feedback Loops, Controlling Filter, and Arousal

It would appear that we are dealing with a number of feedback loops, which at least in part originate and terminate in the same parts of the cortex and which involve, in addition, the striatum, the pallidum, and the thalamus. Moreover, some of the fibers leaving the pallidum appear to innervate the mesencephalic reticular formation rather than the thalamus. The overall impression of this organization is that part of the feedback system deals with precisely targeted functions, whereas others, involving part of the thalamus and the mesencephalic reticular formation, may be engaged in the control of cortical arousal. Taken together, these systems may provide means for the cerebral cortex to control its sensory input by adjusting a thalamic filter, as well as to control the activity of the subcortical arousal systems.

In this feedback mechanism, the mesostriatal dopamine pathways appear to play an important modulatory role. They seem to inhibit the striatum, which in turn is a powerful inhibitory structure, acting, for example, on the thalamus/mesencephalic reticular formation. Stimulation of dopaminergic mechanisms will thus serve to counteract this inhibition, leading to an increased flow of sensory information and arousal. The corticostriatal glutamate system will act in the opposite direction by stimulating the inhibitory function of the striatum. Insofar as the mesencephalic reticular formation is concerned, it is tempting to suggest that the noradrenergic system of the locus ceruleus, and perhaps also the serotonergic system of the raphe nuclei, participate in the control of arousal, mediated via the dorsal and ventral striata. The close relationship between the striatum and the basal nucleus of Meynert, whose cholinergic system appears to promote arousal, is also worth mentioning (Richardson and DeLong 1988).

Lesions or imbalances in the feedback system may lead to different mental and motor disorders. Huntington's chorea provides a good argument for the striatum's playing a role in mental, as well as cognitive, functions. Schizophrenia-like symptomatology may precede the motor disturbances by several years in this neurodegenerative disorder, which is essentially confined to lesions in the striatum (Mattsson 1974). Lesions of the prefrontal cortex, as well as lesions of the pathways from the cortex to the striatum, may produce a "disinhibition syndrome" with an increased sensitivity to amphetamine (Iversen 1977, p. 364). (There are also indications that lesions in this area may cause inability to switch between different activity programs; see the following section.)

The idea of a disturbance in arousal and in a filter function in schizophrenia is, of course, not new. From McGhie and Chapman's (1961) description of the experiences of early schizophrenic patients, Venables (1987) suggested that these patients are "flooded by sensory impressions from all quarters" (p. 204). Lehmann (1966) proposed that if a person has an adequate central processing apparatus to cope with a supernormal influx of stimuli, he will have exceptional creativity. However, if his integrative capacity is insufficient, he may become psychotic. In support of disintegration, Arieti (1966) describes severely psychotic patients who are able to grasp only a small piece of sensory input at a time. None of these authors suggested an anatomical substrate for these mechanisms. However, Stevens (1989) proposed that the dorsal and ventral striata "filter, compress, modulate or 'gate' information from sensory to motor systems in neostriatum and from amygdalahippocampal affect and memory processing regions projecting to limbic striatum en route to thalamus, hypothalamus and frontal lobe" (p. 74).

Finally, it is interesting to note that early work on the mode of action of chlorpromazine focused on the mesencephalic reticular formation (see Bradley 1968). It was proposed that the inhibitory effect of chlorpromazine was due to blockade of excitatory noradrenergic synapses in the mesencephalic reticular formation, thus interfering with arousal. Unfortunately, these studies do not seem to have been extended to more dopamine-specific neuroleptics (Bradley 1986). Thus, the possible influence of dopaminergic mechanisms on the results cannot be determined.

The Possible Role of Glutamate

Theoretically, a deficient corticostriatal glutamatergic function should lead to functional disturbances similar to those caused by dopaminergic agonists, such as amphetamines. This seems actually to be the case. Phencyclidine (PCP, or "angel dust") is now generally recognized as a psychotogenic agent, which is often capable of mimicking schizophrenic symptomatology, perhaps even more faithfully than amphetamines (Angrist 1987; Domino and Luby 1973). PCP is not pure from the pharmacological point of view, but its main target appears to be an ion channel linked to the NMDA receptor, one of the major glutamate receptors (Lodge et al. 1987). Essentially, it may be regarded as a noncompetitive glutamate antagonist.

MK-801 is a more specific NMDA antagonist, acting on the same ion-channel site as PCP (Wong et al. 1986). It is obvious that this agent is a strong candidate for testing our hypothesis that the cerebral cortex via the corticostriatal pathway enhances the inhibitory influence of the striatum on thalamic nuclei as well as on the mesencephalic reticular formation, such as to control the sensory input to the cortex as well as the arousal. From previous work with PCP, it has been suggested that this agent can induce release of catecholamines (Clineschmidt et al. 1982a, 1982b, 1982c), which has been interpreted to mean that the corticostriatal glutamate pathway can inhibit catecholamine release. Thus, it has been suggested that the signs of behavioral stimulation by PCP and MK-801 are mediated via the catecholamines.

However, our hypothesis would suggest a more profound effect of the corticostriatal pathway on the striatum. In fact, the catalepsy and strong psychomotor inhibition induced by eliminating the dopaminergic function would be assumed to be caused by an active inhibition of the striatum, now released from the inhibitory influence of the dopaminergic system. One would thus postulate that even in the absence of dopaminergic function, a psychomotor stimulating influence of MK-801 should persist. To our knowledge, this problem had not been studied; therefore, we have started an investigation in this area (see Carlsson and Carlsson 1989a, 1989b; Carlsson and Svensson 1990).

MK-801 was found to cause a dose-dependent stimulation of locomotor activity in mice, whose stores of catecholamines had been virtually depleted by the combined treatment with reserpine and alpha-methyltyrosine. This was shown in experiments in which the locomotor activity was measured by means of photocells in rectangular cages. It was observed, however, that when an animal came to a corner of the cage it was stuck, apparently because it could not switch from forward locomotion to another movement program. This deficit is reminiscent of the "compulsory approaching syndrome" observed in cats following bilateral caudectomy (Villablanca et al. 1976). Therefore, we started to use rotundas, in which the animals, as a rule, only moved in one and the same direction throughout the experiment. This stimulation of locomotor activity (see Figure 8–2) could not be blocked by the dopamine-receptor antagonists haloperidol or raclopride.

Figure 8–2. Effects of various doses of MK-801 on motor activity in monoamine-depleted mice.

Reserpine (10 mg/kg ip) was administered 18 hours and alpha-methyltyrosine (250 mg/kg ip) 30 minutes prior to the ip MK-treatment. Forward locomotion was registered for 30 minutes, beginning 60 minutes after MK-801 administration. Shown are the means and SEM, $N = 4$. There was a significant correlation between dose and number of meters covered in 30 minutes ($r = 0.6$, $P < 0.01$). (Data from Carlsson and Carlsson 1989b.)

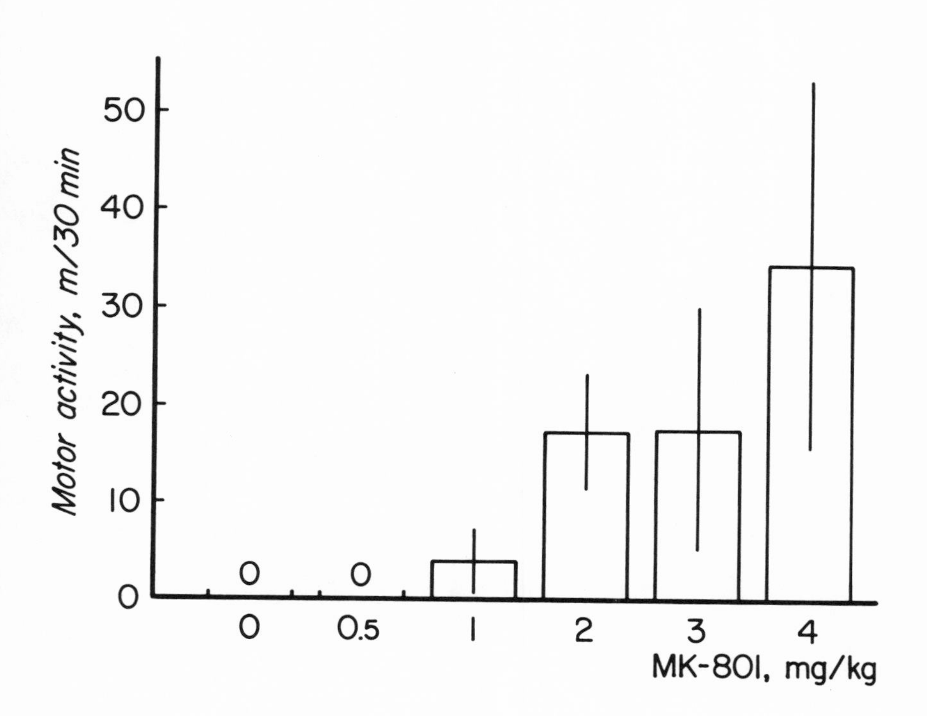

Our experiments have so far been limited to systemic injections of glutamatergic antagonists. However, following striatal injection, NMDA has been found to inhibit and an NMDA antagonist to stimulate psychomotor activity (Schmidt and Bury 1988), a finding that also supports a striatal site of action in our experiments.

We later found that MK-801 is capable of potentiating a variety of arousal-inducing agents, such as clonidine (Figure 8–3), apomorphine, and atropine. The effect of MK-801 plus clonidine could be antagonized by alpha-2-blocking agents but not by an alpha-1-blocker. Interestingly, the stimulat-

Figure 8–3. a) Effects of MK-801 (1 mg/kg ip) and clonidine (2 mg/kg ip), given separately or in combination, on forward locomotion in monoamine-depleted mice; b) an analogous experiment with apomorphine (0.1 mg/kg ip) instead of clonidine. *P* < 0.001, *P* < 0.02 versus MK-801 (Mann-Whitney U test). (Data from Carlsson and Carlsson 1989b.)

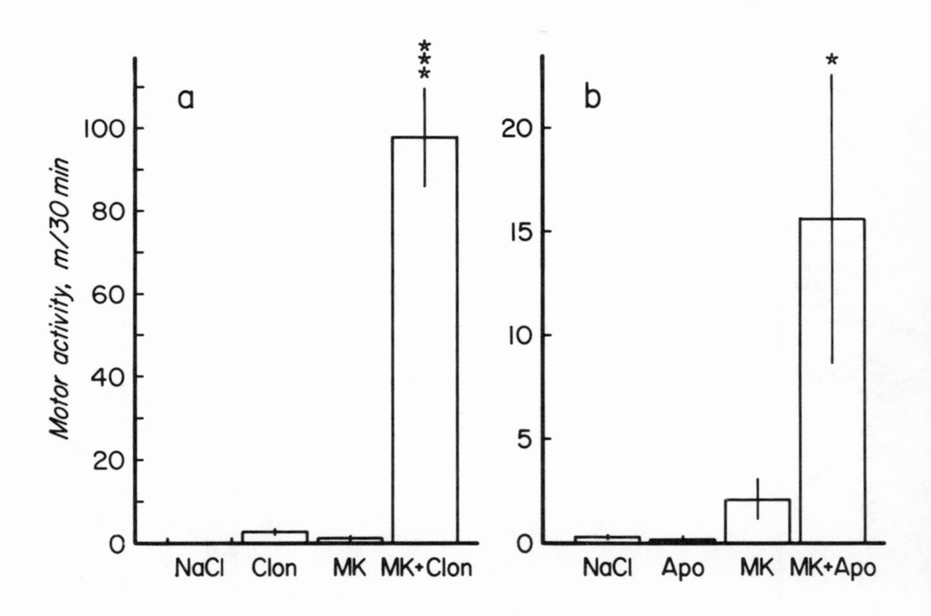

ing action of clonidine could be antagonized by clozapine, which reinforces the notion that this agent may in part exert its antipsychotic action by blocking alpha receptors (see Carlsson 1978).

Summary and Conclusions

We have to revise our views on the role of dopamine in the regulation of mental and motor activities, as well as in the pathogenesis of mental (and perhaps also motor) disorders. Until now, the trend has been to consider dopamine an absolutely essential stimulant for a variety of brain functions: without dopamine, mental and motor activities go down to an extremely low level. Although this is essentially true, it now emerges that the virtual lack of activity in the absence of dopamine or dopamine receptor stimulation is due to an active inhibition exerted via the corticostriatal glutamatergic pathway and striatum on the thalamus and the mesencephalic reticular formation (presumably other subcortical structures are also involved in this regulatory mechanism). If this inhibition is removed or reduced by blocking glutamate receptors, the brain's ability to induce "psychomotor activity" in the absence of dopaminergic stimulation is disclosed. In fact, it appears that the corticostriatal glutamatergic pathway suppressed a number of arousal mechanisms, as indicated, for example, by the activation induced by the alpha-adrenergic agonist clonidine and the antimuscarinic agent atropine in the presence of a subthreshold dose of the NMDA receptor antagonist MK-801. In addition to these postsynaptic mechanisms, the glutamatergic system appears to inhibit the release of catecholamines (see the previous discussion).

The inhibitory action of the corticostriatal glutamate pathway on brain activity may be assumed to be selective in that it allows for adequate and purposeful responses to external stimuli. When the pathway is blocked by MK-801, only forward locomotion appears to be activated at the expense of other programs. This suggests that the selection of purposeful programs, and the switch from one program to another, depends on the inhibitory function of the corticostriatal glutamate pathway.

In many respects the corticostriatal glutamate pathway can be looked upon as an antagonist to the mesostriatal dopaminergic system. Thus, because a psychotic condition can be induced, aggravated, or alleviated by manipulation of the dopaminergic system, it seems reasonable to assume that manipulation of the corticostriatal glutamate pathway can lead to similar consequences. In fact, the psychotogenic action of PCP and ketamine supports this view. Schizophrenia may be induced by a deficiency of the corticostriatal glutamate pathway. Certain observations on postmortem brains of schizophrenic patients do indeed suggest that abnormalities of glutamatergic pathways exist in schizophrenia, even though an effect of chronic neuroleptic treatment remains to be excluded (Kornhuber et al. 1989).

By the same token, glutamatergic agonists may prove to possess antipsychotic properties and to be clinically useful, provided that one can circumvent the side effects, which can be anticipated in view of the ubiquitous occurrence of glutamatergic systems in the brain.

References

Alexander GE, DeLong MR, Strick PL: Parallel organization of functionally segregated circuits linking basal ganglia and cortex. Annu Rev Neurosci 9:357–381, 1986

Angrist B: Pharmacological models of schizophrenia, in Handbook of Schizophrenia, Vol 2. Neurochemistry and Neuropharmacology of Schizophrenia. Edited by Henn FA, Delisi LE. Amsterdam, Elsevier, 1987

Arieti SA: Schizophrenic cognition, in Psychopathology of Schizophrenia. Edited by Hoch PH, Zubin J. New York, Grune & Stratton, 1966

Björklund A, Lindvall O: Catecholaminergic brain stem regulatory systems, in Handbook of Physiolology—The Nervous System, IV. Edited by Field J. Washington, DC, American Physiological Society, 1986

Bradley PB: The pharmacology of synapses in the central nervous system, in Recent Advances in Pharmacology, 4th Edition. Edited by Robson JM, Stacey RS. London, Churchill, 1968

Bradley PB: Pharmacology of antipsychotic drugs, in The Psychopharmacology and Treatment of Schizophrenia. Edited by Bradley PB, Hirsch SR. Oxford, Oxford University Press, 1986

Carlsson A: Antipsychotic drugs, neurotransmitters and schizophrenia. Am J Psychiatry 135:164–173, 1978

Carlsson A: The current status of the dopamine hypothesis of schizophrenia. Neuropsychopharmacology 1:179–186, 1988

Carlsson M, Carlsson A: The NMDA antagonist MK-801 causes marked locomotor stimulation in monoamine depleted mice. J Neural Transm 75:221–226, 1989a

Carlsson M, Carlsson A: Dramatic synergism between MK-801 and clonidine with respect to locomotor stimulatory effect in monoaminedepleted mice. J Neural Transm 77:65–71, 1989b

Carlsson M, Svensson A: Interfering with glutamatergic neurotransmission by means of NMDA antagonist administration discloses the locomotor stimulatory potential of other transmitter systems. Pharmacol Biochem Behav 36:45–50, 1990

Clineschmidt BV, Martin GE, Bunting PR: Anticonvulsant activity (+)-5-methyl-10,11-di-hydro-5H-dibenzo(a,d)cyclohepten-5,10-imine (MK-801), a substance with potent anticonvulsant, central sympathomimetic, and apparent anxiolytic properties. Drug Development Research 2:123–134, 1982a

Clineschmidt BV, Martin GE, Bunting PR, et al: Central sympathomimetic activity of (+)-5-methyl-10,11-dihydro-5H-dibenzo(a,d)cyclohepten-5,10-imine (MK-801), a substance with potent anticonvulsant, central sympathomimetic, and apparent anxiolytic properties. Drug Development Research 2:135–145, 1982b

Clineschmidt BV, Williams M, Witoslawski JJ, et al: Restoration of shock-suppressed behavior with (+)-5-methyl-10,11-dihydro-5H-dibenzo-(a,d)cyclohepten-5,10-imine (MK-801), a substance with potent anticonvulsant, central sympathomimetic, and apparent anxiolytic properties. Drug Development Research 2:147–163, 1982c

Domino E, Luby ED: Abnormal mental states induced by phencyclidine as a model of schizophrenia, in Psychopathology and Psychopharmacology. Edited by Cole JO, Freedman AM, Friedhoff AJ. Baltimore, MD, Johns Hopkins University Press, 1973

Farde L, Pauli S, Hall H, et al: Stereoselective binding of 11C-raclopride—A search for extrastriatal central D2–dopamine receptors by PET. Psychopharmacology (Berlin) 94:471–478, 1988

Goldman-Rakic PS, Selemon LD: Topography of corticostriatal projections in nonhuman primates and implications for functional parcellation of the neostriatum, in Cerebral Cortex, Vol 5. Edited by Jones EG, Peters A. New York, Plenum, 1986

Heimer L, Alheid GF, Zaborszky L: Basal ganglia, in The Rat Nervous System, Vol 1: Forebrain and Midbrain. Edited by Paxinos G. New York, Academic, 1985

Iversen SD: Brain dopamine systems and behavior, in Handbook of Psychopharmacology, Vol 8. Edited by Iversen LL, Iversen SD, Snyder SH. New York, Plenum, 1977, pp 333–384

Kornhuber J, Mack-Burkhardt F, Riederer P, et al: (3-H)MK-801 binding sites in post-mortem brain regions of schizophrenic patients. J Neural Transm 77:231–236, 1989

Lehmann H: Pharmacotherapy of schizophrenia, in Psychopathology of Schizophrenia. Edited by Hoch PH, Zubin J. New York, Grune & Stratton, 1966

Lodge D, Aran JA, Church J, et al: Excitatory amino acids and phencyclidine drugs, in Excitatory Amino Acid Transmission. Edited by Hicks TP, Lodge D, McLennan H. New York, Alan Liss, 1987

Mattsson B: Huntington's chorea in Sweden, II: social and clinical data. Acta Psychiatr Scand Suppl 255:221–235, 1974

McGhie A, Chapman J: Disorders of attention and perception in early schizophrenia. Br J Med Psychol 34:103–110, 1961

Narabayashi H: Lessons from stereotaxic surgery using microelectrode techniques in understanding Parkinsonism. Mt Sinai J Med 55:50–57, 1988

Nauta WJF: Reciprocal links of the corpus striatum with the cerebral cortex and limbic system: a common substrate for movement and thought? in Neurology and Psychiatry: A Meeting of Minds. Edited by Mueller J. Basel, Karger, 1989

Richardson RT, DeLong MR: A reappraisal of the functions of the nucleus basalis of Meynert. Trends Neurosci 11:264–267, 1988

Schmidt WJ, Bury D: Behavioural effects of *N*-methyl-D-aspartate in the anterodorsal striatum of the rat. Life Sci 43:545–549, 1988

Selemon LD, Goldman-Rakic PS: Longitudinal topography and interdigitation of corticostriatal projections in the Rhesus monkey. J Neurosci 5:776–794, 1985

Stevens JR: The search for an anatomic basis of schizophrenia: review and update, in Neurology and Psychiatry: A Meeting of Minds. Edited by Mueller J. Basel, Karger, 1989, pp 64–87

Venables PH: Cognitive and attentional disorders in the development of schizophrenia. in Search for the Causes of Schizophrenia. Edited by Häfner H, Gattaz WF, Janzarik W. Berlin, Springer, 1987, pp 203–213

Villablanca JR, Marcus RJ, Olmsted CE: Effects of caudate nuclei or frontal cortical ablations in cats. Exp Neurol 52:389–420, 1976

Wong EHF, Kemp JA, Priestley T, et al: The anticonvulsant MK-801 is a potent *N*-methyl-D-aspartate antagonist. Proc Natl Acad Sci U S A 83:7104–7108, 1986

9

Clozapine: A Major Advance in the Treatment of Schizophrenia—Clinical and Basic Studies

Herbert Y. Meltzer, M.D.
Gary A. Gudelsky, Ph.D.

Clinical psychiatry, at least as it pertains to the seriously mentally ill, has long been an art and, more recently, a science as well. As a science, clinical psychiatry has its roots in systematic diagnosis and psychopharmacology. As an art, it involves the melding of diagnosis, drug treatment, and psychosocial treatments through interpersonal skills that can bridge the gap between the psychiatrist and troubled patients. When inpatient psychiatry is involved, the interpersonal element looms much larger, because psychiatrist and patient must work together in the context of an ever-changing therapeutic community that has a major shaping influence on the outcome of their efforts. Thomas Detre, in whose honor this chapter is written, is a true 20th century master of the science and art of psychiatry. His genius as a clinician and teacher, exceeded only by his accomplishments as the guiding hand of the most successful research-oriented department of psychiatry in the United States, was a major factor in why one of us (H.Y.M.) chose psychiatry as a profession. The research career that ensued following the shaping influence of only a few weeks' exposure to the charisma of this giant of psychiatry may justly be considered part of his legacy. The Hungarian

The research reported in this chapter was supported in part by USPHS MH 41684, MH 41594, MH 42868, and GCRC MO1RR00080, and grants from the Cleveland, Sawyer, Prentiss, and Scottish Rite Foundations. H.Y.M. is the recipient of a USPHS Research Career Scientist Award MH 47808.

We are grateful to Lee Mason for typing the manuscript.

proverb "the believer is happy; the doubter is wise" aptly describes the weltanschauung with which Dr. Detre empowered his students.

Clinical Studies of Clozapine

The cornerstone of the treatment of schizophrenia has been the many classes of neuroleptic drugs introduced following the serendipitous discovery of chlorpromazine's ability to diminish positive psychotic symptoms: delusions, hallucinations, and thought disorder (Davis et al. 1980). Without question, these agents have made the relatively rapid control of acute psychotic symptoms during the initial or recurrent phases of a psychotic episode possible for many patients with schizophrenia. They have also been effective in the prophylaxis (i.e., prevention of relapse) of patients with schizophrenia (Kane and Lieberman 1987). These effects have contributed to the massive decrease in the number of chronically hospitalized schizophrenic patients, decreased the overall cost of treating this illness, improved the quality of life of patients with schizophrenia, and had secondary beneficial effects on these patients' families.

These positive results, however, have not been universal for patients with schizophrenia, and the significant adverse effects of these agents must be weighed against their clinical benefits. It has been estimated that 10% to 20% of patients with schizophrenia may be considered highly resistant to the antipsychotic effects of typical neuroleptic drugs (Davis et al. 1980). These patients have persistent delusions, hallucinations, and thought disorder, including difficulties in concentration, even if the majority of these individuals would be even more psychotic were they not to receive these agents. Thus, it cannot be said that these patients are nonresponders, only that they are inadequate responders or treatment-resistant. A number of symptoms of schizophrenia have been called "negative" in the sense that they reflect the absence of normal function, such as lack of motivation, withdrawal from social activity, loss of interest in and pleasure from a wide range of activities, motor retardation, and perhaps poverty of thought content (Trimble 1986). It is well recognized that these symptoms are much less responsive to neuroleptic drug treatment than are "positive" symptoms (Goldberg 1985; Meltzer 1985). In this era of deinstitutionalization, when patients with schizophrenia must increasingly be responsible for their own needs, negative symptoms may be as great a burden as positive symptoms in the daily lives of such individuals and their families (Meltzer, in press). Although neuroleptic treatment can partially alleviate negative symptoms (Goldberg 1985; Meltzer 1985), it may also aggravate them by producing extrapyramidal symptoms (EPS) and sedation. Indeed, if current concepts relating negative symptoms to decreased dopaminergic activity in the mesocortical dopaminergic system are correct (MacKay 1980; Meltzer 1989), neuroleptic drugs, which are potent dopamine

D-2 receptor antagonists, might be expected to contribute to the development of, or exacerbate, at least some negative symptoms.

Compliance with neuroleptic treatment has been a major problem since the introduction of these agents, perhaps because of the acute and subacute EPS (i.e., akathesia and parkinsonian symptoms) produced especially by the high potency neuroleptic drugs (van Putten and Marder 1987). Even with anticholinergic treatment, it has often not been possible to eliminate the "drugged" effect, such as stiffness, gait, tremors, posture disturbance, and masked facies, that these drugs produce. Furthermore, some 15% to 25% of patients who take these agents develop some tardive motor syndrome, usually tardive dyskinesia (Casey 1987), which may impair their ability to function normally and contribute to noncompliance and medicolegal problems.

For these reasons, there has been a prolonged search for antipsychotic drugs that produce greater effects on both positive and negative symptoms and for drugs that produce fewer EPS and no tardive dyskinesia or tardive dystonia. Because current concepts of the mechanism of action by which antipsychotic drugs alleviate positive symptoms and produce EPS involve blockade of dopamine D-2 receptors in the mesolimbic and nigrostriatal systems, respectively (Meltzer and Stahl 1976), it is possible that a drug that had an advantage in diminishing EPS might also have a weaker antipsychotic efficacy.

Clozapine, a dibenzazepine antipsychotic drug, was synthesized in the 1960s in a search for a superior antipsychotic with fewer EPS. Early clinical studies established that it did produce fewer EPS but with at least average antipsychotic potency (Angst et al. 1971; Gross and Langer 1970). Clozapine was found to produce agranulocytosis to an extent never previously seen with an antipsychotic drug (Amsler et al. 1977; now known to be 1% to 2% with a peak incidence in the first 18 weeks [Krupp and Barnes 1989]). Because of this side effect, clozapine was withdrawn from the usual course of development and would have been entirely dropped were it not for four factors:

1. Some schizophrenic patients who had responded to clozapine could not be restabilized on other drugs (Gerlach et al. 1974).
2. Some patients found the decrease in EPS very attractive and were willing to risk agranulocytosis.
3. The agranulocytosis was found to be reversible if the drug was stopped promptly (Krupp and Barnes 1989).
4. No reliable reports of tardive dyskinesia appeared despite prolonged use in a large group of patients (Casey 1989).

Clinical experience with clozapine, which indicated that it was effective in patients who did not respond to typical neuroleptic drugs (Gerlach et al.

1974), led to the design of a large-scale multicenter trial of clozapine versus standard neuroleptic agents. Because of the increased risk of agranulocytosis, the U.S. Food and Drug Administration (FDA) Division of Psychopharmacology decided that only patients who had failed to respond to typical antipsychotic drugs could be admitted to this study, because a clear advantage in such patients would presumably balance the risk.

The results of that study (Kane et al. 1988) were encouraging. All patients who entered the trial had failed to respond to at least three neuroleptic drugs (and usually many more) and had been ill an average of 15 years. Patients who failed to respond to a prospective 6-week trial of haloperidol or who were intolerant of haloperidol entered the double-blind phase comparing clozapine and chlorpromazine. Clozapine was found to be superior in decreasing both positive and negative symptoms over the next 6 weeks. Thirty-eight of the 126 (30%) clozapine-treated patients responded compared to 5 out of 136 (4%) of the chlorpromazine-treated patients. EPS of clozapine-treated patients were much milder than those of patients who had been treated with haloperidol plus benztropine. On the basis of this trial, clozapine was introduced into clinical use in the United States, but only for treatment-resistant schizophrenia or patients who are intolerant of typical neuroleptic drugs.

The study of Kane and colleagues (1988) that we have just reviewed is the only controlled study of clozapine in treatment-resistant schizophrenia. However, several retrospective studies of clozapine in treatment-resistant schizophrenia have indicated that the drug might have long-term beneficial effects in such patients (Juul Polvsen et al. 1985; Kuha and Miettinen 1986; Lindstrom 1988). A subsequent prospective study tested clozapine in 51 schizophrenic patients who met previously published criteria for treatment-resistance (Meltzer et al. 1989a). The mean duration of treatment was 10.3 ± SD 8.1 months (median 7.6 months). Significant improvement in total Brief Psychiatric Rating Scale (BPRS; Overall and Gorham 1962) score was noted at 6 weeks, with further improvement first occurring at 9 months in the group as a whole. Most of the improvement occurred in the first 6 weeks in the group as a whole. Improvement was noted in both BPRS positive and negative symptoms; analysis of covariance indicated that improvement in negative symptoms was independent of improvement in positive symptoms.

The time course of the achievement of the criteria for improvement in total BPRS score (≥ 20% of baseline) required by the FDA to designate clozapine-treated patients as responders was of particular interest. Thirty-one of the 38 patients who remained on clozapine were considered to be responders by these criteria. The other 7 patients improved in social function and decreased need for hospitalization. Of these 31, 14 (45.2%) responded by 6 weeks, another 9 (29.0%) by 3 months, another 2 (6.5%) at 6 months, another 5 (16.1%) at 9 months, and the last 1 (3.2%) at 12 months. Thus, it was concluded that at least a 6-month clinical trial (and perhaps as much as 12 months) is not unreasonable for patients who fail to respond to other neuroleptic drugs. Clozapine also dramatically improved patients' quality of life as measured by change in social function, work, interpersonal relations, and decreased need for hospitalization (Meltzer et al. 1989a, 1990).

A very low incidence of EPS, along with a blockade of symptoms of tardive dyskinesia by clozapine, has been found in most patients. These results indicate that it is possible to develop a drug with greater efficacy in positive and negative symptoms, and, moreover, that the drug can have additional advantages in producing fewer EPS and no tardive dyskinesia. Clearly, clozapine would be the drug of choice in schizophrenia if it did not produce a high incidence of agranulocytosis. It might also be of advantage for the treatment of affective psychoses and even for indications such as chronic, severe anxiety states that fail to respond to anxiolytic drugs. The risk of tardive dyskinesia with typical neuroleptics has clearly limited the use of such drugs for many possible indications. Therefore, the search for the mechanism of action of clozapine takes on great practical and theoretical significance. Understanding the etiology of psychosis, if not schizophrenia, the relationship between positive and negative symptoms, and the cause of tardive dyskinesia may come from the further study of the action of clozapine.

Preclinical Studies

The preclinical pharmacology of clozapine also differs from that of typical antipsychotic agents. Differences in the actions of clozapine and typical antipsychotic drugs have been reported in behavioral, electrophysiological, biochemical, and neuroendocrinological studies. It is beyond the scope of this chapter to review the entire preclinical pharmacology of clozapine. In the discussion that follows, we focus on actions of clozapine on forebrain dopamine neurons and pituitary hormone secretion, which distinguish this antipsychotic from typical neuroleptics.

Effects on Mesotelencephalic Dopamine Neurons

The acute administration of typical antipsychotic drugs, such as haloperidol, increases the firing rate of dopamine cells of the mesocorticolimbic (A10) and nigrostriatal (A9) pathways (Bunney et al. 1973; Hand et al. 1987; White and Wang 1983), as well as the synthesis and metabolism of dopamine within the terminals of these neurons (Andén 1972; Carlsson 1975; Roth et al. 1976). Results from studies using in vivo dialysis or in vivo voltammetry are also consistent with the view that typical antipsychotic drugs acutely increase the release of dopamine from A9 and A10 dopaminergic neurons (Imperato and DiChiara 1985; Lane and Blaha 1986).

In contrast to the actions of typical antipsychotics on A9 and A10 dopamine neurons, some evidence suggests that clozapine acts selectively on dopamine neurons of the mesocorticolimbic pathway. Consistent with this contention are the findings of Huff and Adams (1980) and Blaha and Lane (1987), who reported that clozapine increased the efflux of dopamine in the nucleus accumbens but not in the striatum, as ascertained by in vivo voltammetry. Moreover, Hand and colleagues (1987) have reported that clozapine increases the firing rate of A10 but not of A9 dopamine neurons. The findings that typical antipsychotics acutely increase the concentrations of 3-methoxytyramine in the striatum, whereas clozapine and other purported atypical antipsychotics do not elicit this effect, further supports the view that clozapine differs from typical neuroleptics in its action on striatal dopamine release (Altar et al. 1988).

However, data from biochemical studies are not entirely supportive of the hypothesis that clozapine selectively activates mesocorticolimbic dopamine neurons. Like haloperidol, clozapine increases the activity of tyrosine hydroxylase in the striatum, as well as in the nucleus accumbens and increases the concentration of DOPAC in both brain regions (Wilk et al. 1975). Moreover, Coward and colleagues (1989) have reported that clozapine does enhance the efflux of striatal dopamine, as measured by in vivo dialysis. In general, it appears that clozapine may enhance the synthesis and metabolism of dopamine within nigrostriatal neurons at doses that do not appreciably alter either the release of dopamine from these neurons or their firing rate. The significance of this dissociation, if real, remains to be examined.

A selective action of clozapine on the function of mesocorticolimbic dopamine neurons has been reported after its repeated administration. The repeated administration of haloperidol has been reported to produce depolarization blockade of A9 and A10 dopamine neurons and suppress the release of dopamine from terminals in the striatum and nucleus accumbens (Blaha and Lane 1987; Chiodo and Bunney 1983; White and Wang 1983). However, depolarization blockade following clozapine was reported to occur only in A10 dopamine cells (Chiodo and Bunney 1983; Hand et al. 1987), and dopamine release was reported to be suppressed only in the nucleus accumbens following the repeated administration of clozapine (Blaha and Lane 1987). Recently, we have found that 21 days of treatment with clozapine produced no significant decrease in dopamine release in either the striatum or nucleus accumbens of rats, whereas chronic haloperidol markedly reduced DA release in both regions. Both drugs increased dopamine release from basal levels, even after repeated administration. The differences between these results and those obtained with electrochemical or electrophysiological methods requires further study but could be due to the use of anesthesia in the former studies. If dopamine release is not decreased in the striatum or nucleus accumbens, this could explain clozapine's lack of EPS and should have some bearing on its antipsychotic activity as well. Chronic clozapine has been found not to decrease DA release in the nucleus accumbens or striatum, whereas typical neuroleptics such as haloperidol decrease DA release in both regions (Ichikawa and Meltzer 1990). It may be that chronic neuroleptic treatment leads to too great an interference with dopaminergic function in the mesolimbic system, whereas clozapine leaves it in a more normal range.

Effects on Hypothalamic Dopamine Neurons and Hormone Secretion

The actions of clozapine and other purported atypical antipsychotics also differ from those of typical antipsychotics on dopamine neurons of the tuberoinfundibular pathway. The acute effects of typical antipsychotic drugs on the activity of nigrostriatal and mesocorticolimbic dopaminergic neurons are generally thought to be mediated by neuronal feedback and/or autoreceptor mechanisms. However, the activity of tuberoinfundibular dopamine (TIDA) neurons is not regulated by neuronal feedback or autoreceptor mechanisms but, rather, by circulating prolactin (Gudelsky 1981; Moore and Demarest 1982). Consequently, the acute administration of haloperidol and other typical antipsychotic drugs does not alter the activity of TIDA neurons (Demarest and Moore 1979; Moore and Demarest 1982). In contrast, clozapine and several other purported atypical antipsychotics (e.g., thioridazine, melperone, RMI 81582) do acutely increase the activity of these hypothalamic dopamine neurons, as judged from the activity of tyrosine hydroxylase in the median eminence (Gudelsky and Meltzer 1989; Gudelsky et al. 1987).

Inasmuch as TIDA neurons inhibit the secretion of prolactin, it can be envisioned that the stimulatory effect of clozapine on TIDA neurons accounts for the finding that plasma prolactin concentrations in patients maintained on clozapine are not significantly elevated (Kane et al. 1981; Meltzer et al. 1979). This finding is in contrast to the sustained elevation of plasma prolactin concentration in patients maintained on typical antipsychotic agents (Meltzer and Fang 1976).

Although, in the rat, acute administration of clozapine produces an increase in serum prolactin concentrations (Gudelsky et al. 1987; Meltzer et al. 1975), the duration of the increase is much shorter than that produced by typical antipsychotics (Gudelsky et al. 1987). Antagonism of the clozapine-induced activation of TIDA neurons by treatment with para-chlorophenylalanine (PCPA) results in an augmentation of the clozapine-induced elevation of serum prolactin concentrations in the rat (S. Berry, personal communication, December 1990). Thus, the stimulatory effect of clozapine on TIDA neurons appears to limit the extent of prolactin secretion. However, under these conditions, the duration of the elevated serum prolactin concentrations produced by clozapine is still relatively brief compared to that produced by haloperidol.

Clozapine and other atypical antipsychotic agents also differ from typical neuroleptics with regard to their effects on the hypothalamic-pituitary-adrenal axis. Plasma concentrations of corticosterone and adrenocorticotropic hormone (ACTH) are relatively unaffected by typical antipsychotics, whereas clozapine, melperone, and RMI 81582 produce dose-related increases in plasma corticosterone and ACTH concentrations in rodents (Gudelsky et al. 1989).

Role of D-1 Receptors

Andersen and Braestrup (1986) have suggested that preferential blockade of D-1 receptors accounts for the unique clinical profile of clozapine. There is evidence that the ratio of affinities of atypical antipsychotics for D-1 and D-2 receptors differs from that for typical antipsychotics (Coward et al. 1989; Meltzer et al. 1989b).

There is also some evidence for the role of D-1 receptors in the actions of clozapine (Coward et al. 1989). A D-1 agonist, CY 208-243, completely suppresses the clozapine-induced release of dopamine in the striatum. Coward and colleagues (1989) have suggested that, in small doses (e.g., 5 mg/kg), clozapine produces a selective blockade of D-1 receptors in the rat, whereas at large doses (e.g., 20 mg/kg) clozapine produces blockade of both D-1 and D-2 receptors.

D-1 receptor activation also modifies the actions of clozapine on TIDA neurons. The clozapine-induced activation of tyrosine hydroxylase in TIDA neurons is completely suppressed by the D-1 agonist SKF 38393 (Gudelsky and Meltzer 1989). However, simple blockade of D-1 receptors by clozapine does not appear to be the mechanism by which clozapine activates TIDA neurons because the selective D-1 antagonist SCH 23390 does not increase the activity of tyrosine hydroxylase in TIDA neurons. Moreover, D-1 receptor stimulation prevents or reverses the activation of TIDA neurons induced by a number of different treatments, such as reserpine, neurotensin, and hyperprolactinemia (Berry et al. 1990). Thus, D-1 agonists do not selectively suppress the clozapine-induced activation of tyrosine hydroxylase in TIDA neurons.

Role of Serotonergic Mechanisms

Considerable evidence has shown that clozapine and other purported atypical antipsychotics differ from typical neuroleptics in their interactions with

5-HT receptors and neuronal systems. Although many antipsychotic drugs appear to bind appreciably to some 5-HT receptor subtypes, as well as to dopamine receptors, it has been suggested that both 5-HT_2 and dopaminergic receptors are involved in the mechanism of action of antipsychotic drugs. In addition, it has been reported that the use of antipsychotics that possess significant 5-HT_2 antagonist properties is associated with a low incidence of extrapyramidal effects (Altar et al. 1986; Meltzer et al. 1989b).

Indeed, atypical antipsychotics appear to be more effective 5-HT antagonists than typical neuroleptics. Clozapine has been shown to block fenfluramine-induced hyperthermia (Sulpizio et al. 1978) and the 5-HT related effect of LSD to potentiate apomorphine-induced hypermotility (Fink et al. 1984). Fuller and Mason (1986) reported that clozapine, but not fluphenazine, attenuated quipazine-induced corticosterone secretion. In addition, Nash and colleagues (1988) have shown that atypical, but not typical, neuroleptics attenuate 5-HT_2 receptor-mediated changes in body temperature and corticosterone secretion.

These findings of the functional effects of atypical antipsychotics at 5-HT_2 receptors in vivo are consistent with data from radioligand binding studies. Altar and colleagues (1986) reported that several atypical antipsychotics, including clozapine, differed from typical antipsychotics in their relative affinities for 5-HT_2 and D-2 receptors. That is to say, clozapine-like drugs displayed a greater affinity for 5-HT_2 binding sites and a lower affinity for D-2 sites as compared to typical neuroleptics. Meltzer and colleagues (1989b) have analyzed the pKi values (negative log of affinities) of numerous typical and purported atypical antipsychotics for 5-HT_2, D-1, and D-2 binding sites. Atypical antipsychotics were found to have lower pKi values for the D-2 receptor than typical neuroleptics. In addition, 90% of the tested antipsychotics could be classified correctly as typical or atypical on the basis of the ratio of pKi 5-HT_2/pKi D-2. Consistent with the findings of Altar and colleagues (1986), Meltzer and colleagues (1989b) reported that atypical antipsychotics, more so than typical antipsychotics, have a relatively greater affinity for 5-HT_2 receptors than for D-2 receptors.

Serotonergic mechanisms also appear to be involved in the stimulatory effect of clozapine on TIDA neurons. In rats in which brain 5-HT has been depleted by PCPA or 5-7-dihydroxytryptamine (5,7-DHT), the stimulatory effect of clozapine on tyrosine hydroxylase activity in the median eminence is markedly suppressed (S. Berry, personal communication, December 1990). Thus, it would appear that 5-HT neurons are required for the stimulatory effects of clozapine on TIDA neurons. On the basis of this finding, it is reasonable to suggest that the stimulatory effect of clozapine on TIDA neurons involves an activation of serotonergic neurons. However, it should be noted that Gallagher and Aghajanian (1976) have shown that clozapine, unlike typical neuroleptics, inhibits the firing rate of 5-HT neurons in the raphe nucleus. Thus, the exact nature of the dependence on intact 5-HT neurons for the effects of clozapine on TIDA neurons is unclear.

The stimulatory effects of clozapine on forebrain dopamine systems also appears to depend on intact 5-HT neurons. The clozapine-induced increase in DOPAC concentrations in the medial prefrontal cortex and nucleus accumbens is greatly attenuated in rats in which brain 5-HT has been depleted by PCPA or 5,7-DHT. Intact 5-HT neurons appear to be important for the stimulatory effect of clozapine on mesocorticolimbic but not nigrostriatal dopamine neurons, because the clozapine-induced increase in striatal DOPAC concentrations is not diminished in rats treated with PCPA or 5,7-DHT (J. Brodkin, personal communication, October 1990).

It is interesting to note that the effects of haloperidol, like those of clozapine, on DOPAC concentrations in the medial prefrontal cortex are attenuated in rats depleted of 5-HT, whereas the effects in the striatum are unaltered. The stimulatory effect of haloperidol on DOPAC concentrations in the nucleus accumbens is not as dependent upon intact 5-HT neurons as is the effect of clozapine. Thus, in certain brain regions, interactions between 5-HT and DA may be more important for the atypical antipsychotics than for the typical neuroleptics.

Summary

Strong evidence shows that clozapine is the first antipsychotic drug to decrease positive and negative symptoms in schizophrenic patients more effectively than other antipsychotic drugs. This alone makes the drug a major milestone in the history of schizophrenia. That clozapine also does not produce tardive dyskinesia is of immense consequence. Were it not to produce agranulocytosis in 1% to 2% of cases, clozapine would be the drug of choice for all schizophrenic patients and open up the use of antipsychotic drugs in many other conditions where psychotropic drugs have been of value but their use limited because of EPS, especially tardive dyskinesia.

It is reasonable to hope that other clozapine-like drugs that do not produce agranulocytosis will be discovered. It falls to basic research to clarify the mechanism of action of clozapine in order to identify candidate antipsychotic drugs with clozapine-like properties. Although intensive investigations of clozapine have been conducted since it was first discovered 20 years ago, no definitive explanation of its unique properties exists. It does not appear that effects on dopamine are important, but its mechanism of action cannot simply be blockade of D-2 dopamine receptors. Evidence suggests that 5-HT_2 blockade (Altar et al. 1986; Meltzer 1989) may be of great importance, but it is likely that other factors are also important. As the mechanism of action of clozapine is clarified, our understanding of the etiology of schizophrenia and tardive dyskinesia is also likely to increase.

References

Altar CA, Wasley AM, Neale RF, et al: Typical and atypical antipsychotic occupancy of D_2 and S_2 receptors: an autoradiographic analysis in rat brain. Brain Res Bull 16:517, 1986

Altar CA, Boyar WC, Wasley A, et al: Dopamine neurochemical profile of atypical antipsychotics resembles that of D_1 antagonists. Naunyn Schmiedebergs Arch Pharmacol 338:162–168, 1988

Amsler HA, Teerenhovi L, Barth E, et al: Agranulocytosis in patients treated with clozapine. Acta Psychiatr Scand 56:241–248, 1977

Andén N-E: Dopamine turnover in the corpus striatum and the limbic system after treatment with neuroleptic and anti-acetylcholine drugs. Journal of Pharmacy and Pharmacology 24:905, 1972

Andersen PH, Braestrup C: Evidence for different states of the dopamine D_1 receptor: clozapine and fluperlapine may preferentially label an adenylate cyclase-coupled state of the D_1 receptor. J Neurochem 47:1822–1831, 1986

Angst J, Bente D, Berner P, et al: Dasklinische werkungskild von clozapine (Untersuchung mid deur AMP-Syst). Pharmakopsychiatrie 4:200–211, 1971

Berry SA, Meltzer HY, Gudelsky GA: D_1 receptor stimulation inhibits the activation of tuberoinfundibular dopamine neurons. J Pharmacol Exp Ther 254:677–682, 1990

Blaha CD, Lane RF: Chronic treatment with classical and atypical antipsychotics drugs differentially decreases dopamine release in striatum and nucleus accumbens *in vivo*. Neurosci Lett 78:188–204, 1987

Bunney BA, Walters JR, Roth RH, et al: Dopaminergic neurons: effects of antipsychotic drugs and amphetamine on single cell activity. J Pharmacol Exp Ther 185:560–572, 1973

Carlsson A: Receptor-mediated control of dopamine metabolism, in Pre- and Postsynaptic Receptors. Edited by Usdin E and Bunney WE Jr. New York, Marcel Dekker, 1975

Casey DE: Tardive dyskinesia, in Psychopharmacology: The Third Generation of Progress. Edited by Meltzer HY. New York, Raven, 1987

Casey DE: Clozapine: Neuroleptic-induced EPS and tardive dyskinesia. Psychopharmacology 99:S47–S53, 1989

Chiodo LA, Bunney BS: Typical and atypical neuroleptics: differential effects of chronic administration on the activity of A9 and A10 midbrain dopaminergic neurons. J Neurosci 3:1607–1619, 1983

Coward DM, Imperato A, Urwyler S, et al: Biochemical and behavioral properties of clozapine. Psychopharmacology 99:S6–S12, 1989

Davis JM, Schaffer CB, Killian GA, et al: Important drug issues in the treatment of schizophrenia. Schizophr Bull 6:70–87, 1980

Demarest KT, Moore KE: Comparison of dopamine synthesis regulation in the terminals of nigrostriatal, mesolimbic, tuberoinfundibular and tuberohypophyseal neurons. J Neural Transm 46:263–277, 1979

Fink H, Morgenstern R, Oelssner W: Clozapine—a serotonin antagonist? Pharmacol Biochem Behav 20:513, 1984

Fuller RW, Mason NR: Flumezapine, an antagonist of central dopamine and serotonin receptors. Res Commun Chem Pathol Pharmacol 54:23, 1986

Gallagher DW, Aghajanian GK: Effect of antipsychotic drugs on the firing of dorsal raphé cells, I: role of adrenergic system. Eur J Pharmacol 39:341–355, 1976

Gerlach J, Kopplehaus E, Helweg E, et al: Clozapine and haloperidol in a single-blind cross-over trial: therapeutic and biochemical aspects in the treatment of schizophrenia. Acta Psychiatr Scand 50:410–424, 1974

Goldberg S: Negative and deficit symptoms in schizophrenia do respond to neuroleptics. Schizophr Bull 11:453–456, 1985

Gross H, Langer H: Das neuroleptikum 100-129/#F-1854 (clozapine) in der psychiatrie. International Pharmakopsychiatrie 4:220–230, 1970

Gudelsky GA: Tuberoinfundibular dopamine neurons and the regulation of prolactin secretion. Psychoneuroendocrinology 6:3–16, 1981

Gudelsky GA, Meltzer HY: Activation of tuberoinfundibular dopamine neurons following the acute administration of atypical antipsychotics. Neuropsychopharmacology 2:45–51, 1989

Gudelsky GA, Koenig JI, Simonovic M, et al: Differential effects of clozapine and haloperidol on tuberoinfundibular dopamine neurons and prolactin secretion in the rat. J Neural Transm 68:227–240, 1987

Gudelsky GA, Nash JF, Berry SA, et al: Basic biology of clozapine: electrophysiological and neuroendocrinological studies. Psychopharmacology 99:S13–S17, 1989

Hand TH, Hu X-T, Wang RY: Differential effects of acuteclozapine and haloperidol on the activity of ventral tegmenta (A10) and nigrostriatal (A9) dopamine neurons. Brain Res 415:257–269, 1987

Huff R, Adams RN: Dopamine release in n. accumbens and striatum by clozapine: simultaneous monitoring by *in vivo* electrochemistry. Neuropharmacology 19:587–590, 1980

Ichikawa J, Meltzer HY: The effect of chronic clozapine and haloperidol on basal dopamine release and metabolism in rat striatum and nucleus accumbens studied by *in vivo* dialysis. Eur J Pharmacol 176:371–374, 1990

Imperato A, DiChiara G: Dopamine release and metabolism in awake rats after systemic neuroleptics as studied by trans-striatal dialysis. J Neurosci 5:297–306, 1985

Juul Polvsen U, Noring U, Fog R, et al: Tolerability and therapeutic effect of clozapine: a retrospective investigation of 216 patients treated with clozapine for up to 12 years. Acta Psychiatr Scand 71:176–185, 1985

Kane JM, Lieberman JN: Maintenance pharmacotherapy in schizophrenia, in Psychopharmacology: The Third Generation of Progress. Edited by Meltzer HY. New York, Raven, 1987

Kane J, Cooper TB, Sachar EJ, et al: Clozapine: plasma levels and prolactin response. Psychopharmacology 73:184–187, 1981

Kane J, Honigfeld G, Singer J, et al: The clozaril collaborative study group: clozapine for the treatment-resistant schizophrenic: a double-blind comparison with chlorpromazine. Arch Gen Psychiatry 45:789–796, 1988

Krupp P, Barnes P: Leponex—associated granulocytopenia: a review of the situation. Psychopharmacology 99:S118–S121, 1989

Kuha S, Miettinen E: Long-term effect of clozapine in schizophrenia: a retrospective study of 108 chronic schizophrenics treated with clozapine for up to 7 years. Nord Psychiatr Tidskr 40:225–230, 1986

Lane RF, Blaha CD: Electrochemistry in vivo: application to CNS pharmacology. Ann N Y Acad Sci 473:50–69, 1986

Lindstrom LH: The effect of long-term treatment with clozapine in schizophrenia: a retrospective study in 96 patients treated with clozapine for up to 13 years. Acta Psychiatr Scand 77:524–529, 1988

MacKay AVP: Positive and negative symptoms and the role of dopamine. Br J Psychiatry 137:378–386, 1980

Meltzer HY: Dopamine and negative symptoms in schizophrenia: critique of the Type I-Type II hypothesis, in Controversies in Schizophrenia: Changes and Constancies. Edited by Alpert M. New York, Guilford, 1985

Meltzer HY: Clinical studies on the mechanism of action of clozapine: the dopamine-serotonin hypothesis of schizophrenia. Psychopharmacology 99:S18–S27, 1989

Meltzer HY: Pharmacological treatment of negative symptoms, in Negative Schizophrenic Symptoms: Pathophysiology and Clinical Implications. Edited by Greden JF, Tandy R. New York, Spectrum Press (in press)

Meltzer HY, Fang V: The effect of neuroleptics on serum prolactin levels in schizophrenic patients. Arch Gen Psychiatry 30:279–286, 1976

Meltzer HY, Stahl SM: The dopamine hypothesis of schizophrenia: a review. Schizophr Bull 2:19–76, 1976

Meltzer HY, Daniels S, Fang VS: Clozapine increases rat serum prolactin levels. Life Sci 17:339–342, 1975

Meltzer HY, Goode DJ, Schyve PM, et al: Effect of clozapine on human serum prolactin levels. Am J Psychiatry 136:1550–1555, 1979

Meltzer HY, Bastani B, Kwon KY, et al: A prospective study of clozapine in treatment-resistant patients, I: preliminary report. Psychopharmacology 99:S68–S72, 1989a

Meltzer HY, Matsubara S, Lee J-C: Classification of typical and atypical antipsychotic drugs on the basis of dopamine D-1, D-2 and serotonin$_2$ pKi values. J Pharmacol Exp Ther 251:238–246, 1989b

Meltzer HY, Burnett S, Bastani B, et al: Effect of six months of clozapine treatment on the quality of life of chronic schizophrenic patients. Hosp Community Psychiatry 41:892–897, 1990

Moore KE, Demarest KT: Tuberoinfundibular and tuberohypophyseal dopaminergic neurons, in Frontiers in Neuroendocrinology, Vol 7. Edited by Ganong WF, Martini L. New York, Raven, 1982

Nash JF, Meltzer HY, Gudelsky GA: Antagonism of serotonin receptor-mediated neuroendocrine and temperature responses by atypical neuroleptics in the rat. Eur J Pharmacology 151:463–469, 1988

Overall JE, Gorham DR: Brief Psychiatric Rating Scale (BPRS). Psychol Rep 10:799–812, 1962

Roth RH, Murrin LC, Walters JR: Central dopaminergic neurons: effects of alterations in impulse flow on the accumulation of dihydroxyphenylacetic acid. Eur J Pharmacology 36:163–172, 1976

Sulpizio A, Fowler PJ, Macko E: Antagonism of fenfluramine-induced hyperthermia: a measure of central serotonin inhibition. Life Sci 22:1439–1446, 1978

Trimble MR: Positive and negative symptoms in schizophrenia. Br J Psychiatry 148:587–589, 1986

Van Putten T, Marder SR: Behavioral toxicity of antipsychotic drugs. J Clin Psychiatry 48 (suppl):13–19, 1987

White FJ, Wang RY: Differential effects of classical and atypical antipsychotic drugs on A9 and A10 dopamine cells. Science 221:1054–1057, 1983

Wilk S, Watson E, Stanley ME: Differential sensitivity of two dopaminergic structures in rat brain to haloperidol and to clozapine. J Pharmacol Exp Ther 195:265–270, 1975

10

Summarizing Comments: The Past Predicts the Future

Daniel X. Freedman, M.D.

The symposium on which this volume is based was, in part, a celebration of the sweep and promise of modern science. For the bedazzled clinician, the reach of available and emerging investigative approaches almost exceeds our capacity to gasp. If we are to be appreciative but not mesmerized by what may soon be in hand, we will have to determine which of the plethora of basic science leads to follow in the clinic and when to do so. This challenge is problematic for all of medicine. Nevertheless, with drugs to change and probe basic biobehavioral processes, as well as to treat disorders, and with psychometric tools and clinical research designs to objectify the characteristics of disease, we surely have the right to enjoy the current momentum of therapeutic and pathophysiologic research in psychiatry. Another point is also clear: if one has to be ill—and we now soundly know that some of us do—one is far better off being so today than 40 or more years ago. There is reason, then, for celebration—and for reflection.

Accordingly, it is worth recalling the way things were with science. In that sense, the past held its wonders but knowingly wandered in our current directions as well. Certainly, one can imagine the sheer excitement and astonishment in establishing not only the etiology of but the power of prevention of psychiatry's "two Ps"—pellagra and paresis. Imagine, too, the newborn confidence and hope of those who discovered that electroconvulsive therapy (ECT) terminated depression, shifting the therapeutic task from a grim endurance contest of watchful waiting while the deep mourning of depression ran its painful course, to a process of active intervention to speed recovery.

What is "Progress" in Psychiatric Therapeutics?

Table 10–1 telegraphically compresses this larger perspective of probings and progress with respect to the treatment of major disabling psychiatric disorders. From that broad sweep of history (Redlich and Freedman 1966; Valenstein 1986), there are three points I would emphasize.

First, at the beginning of the century, psychiatry was using combined moral or supportive therapies and any available somatic therapy that held

Table 10–1. Treatment of psychoses—past, present, and future

c. 1900	Nonsomatic "moral" therapies: asylum from stress; re-education Somatic therapies: chloral, opiates, bromides, and barbiturates; electrotherapy; banotherapy; rare trephinings
1920–1940	Drug and electrically induced sleep; dental, gastrointestinal, genital, and endocrine surgeries ("local infections"); CO_2 inhalation; hyperbaric O_2; stimulants-sedatives; immunotherapy (antigens in CSF); *malaria fever therapy** Metrazol convulsions; ECT; insulin coma; first *lobotomies**
1940–1950	Penicillin; "total push" and occupational therapies; amytal narcosynthesis; stimulants, psychoanalytic and "direct" treatments for psychosis
1952–1966	Reserpine; neuroleptics; monoamine oxidase inhibitors; tricyclics; lithium; classic and new benzodiazepines; psychosocial rehabilitation and group and family therapies—move to outpatient departments and general hospital therapies
1966–present	Secondary prevention (earlier drug and maintenance therapies); rehabilitation and behavior therapies; brief psychotherapies; new uses of old drugs (clonidine, carbamazepine); available drugs with semiselective receptor profiles for affective, anxiety, and obsessive-compulsive disorders
In the near future	Drug design: based on brain regional targets and receptor and membrane channel effects; aimed at disease subtypes or behavioral dimensions (reactivity) or side-effect profiles; "predictors" for drug choices Chronically ill: symptomatically selective drugs (for negative symptoms) and match of personality and environment
2000+	Basis for therapy: gene products and expression; psychoneurobiology of disease mechanisms: "ligandomimetics," based on nature's logic

* Nobel Prize.

faint promise or, for various theoretical reasons, allure. We are still doing just that.

The period of drug discovery between 1952 and 1966—as in all of medicine—took time to permeate the various infrastructures of the profession. Yet, if we think about the subsequent years (our recent past), we have mainly been fine-tuning what we have long had in hand. More recently, receptor selective drugs and unexpected uses for them (for example, 5-HT_{1a} drugs for anxiety and depression) have rightly captured attention. But, essentially, having found drugs with efficacy, we have been freer to shift our focus to secondary prevention—maintenance therapies and the like.

In the cleverly designed but labor intensive work with the problems of relapse in depressive disorder that Kupfer and colleagues have been pursuing (Frank et al. 1990), combined therapies indeed seem to have a role. Clearly, the notion that psychotherapy can be dispensed with is not sustained by this systematic work. Rather, psychotherapy (specifically, interpersonal psychotherapy, or IPT) promises some aid in preventing relapse or, in combination with drug therapy, of shortening the duration of the relapse episode as compared to the carefully studied index episodes. Thus, my first and major point is that, even with the best of our drugs, we are not likely to be doing without psychotherapies in combination with somatic therapies for the affective and anxiety disorders. Precision in specification of both is a challenge; so too for the retraining and rehabilitative treatments of the schizophrenias (where, as Meltzer and Gudelsky emphasize in Chapter 9, new drug approaches can be anticipated).

My second major moral, reflected in Table 10–1, is that the pharmacological discoveries from 1952 to 1966 have not given us disease-specific drugs. They are, at best, symptomatic treatments, something approaching cimetidine for gastric ulcers where, even with the pathophysiology in hand, we still do not know the etiopathology. Thus, what we can envisage as likely is that we will find predictors in the future—predictors for choice of therapies, even though etiopathologic specificities such as for our "2 Ps" elude us. We will continue to pull drugs from the shelf and out of the pipeline and stumble ahead (or make brief, brilliant runs) to find new uses and effects—as with serotonergic drugs for obsessive-compulsive disorders. Rarely (to continue a football analogy), we may cross the goal line and score.

We, unarguably, have a variety of therapeutic options. This is a fundamental fact of the moment and should be taken seriously. One can interpret many of the contributions to this volume as at least suggesting that the reason we have these options is not only ignorance. Rather, it may rest upon the way physiologic regulators work and how the brain is built. Most of therapeutics is compensatory; and, in the establishment of new equilibria, it is conceivable that the brain's intrinsic accessibility to (and need for) different paths of influence renders the very existence of therapeutic options inevitable.

Fundamentally, what we really seek is to use nature's logic to develop drugs with as much specificity as the system will allow. Accordingly, perhaps by the year 2000, we will have what I call "ligandomimetics" that are thoroughly based on such logic. Further, this approach may help in dissecting the various ways by which invisible brain operations that are manifest in overt phenomenology are fundamentally dissociable and manipulable. This happened for Freud and others using accidents of nature to infer brain operations from the phenomena of the aphasias and agnosias. If we make the ideal case out of the recent observations that intractable obsessive thoughts lose their preemptive grip on the patient's attention and behavioral repertoire with the use of drugs affecting a specific subtype of serotonergic activity, we can observe such inferential steps. And we may next learn from positron-emission tomography (PET) scans, where, with symptomatic improvement, the differential hypermetabolism of the brain's frontal and basal ganglia regions shows distinct shifts. With a patient's recovery, hypermetabolism diminishes. Perhaps the frontal lobes, in anticipating and dealing with the environment (and, in doing so, permitting appropriate balance between proximal and distal concerns), are less overworked in the recovered patient. They now are freed to return to their task; concomitantly, the basal ganglia resume their phylogenetically defined role in implementing routine and ritual rather than shifting such functions to frontal regions (Baxter et al. 1987). Whether or not this fantasy is reasonably thinkable, we may certainly anticipate both precision and surprise as we continue to use the advances of current science to grasp how behaviors are put together.

This leads me to a third moral. In all of medicine, therapies generally precede our knowledge of mechanisms—certainly of mechanisms of drug action that, in turn, generally precede our knowledge of mechanisms of drug effect (because that is always more complicated). Penicillin's molecular mode of action on bacterial membranes was not known until 10 to 12 years after the drug began to be used. Even as we can now sometimes begin with remarkable initial knowledge of molecular modes of action, we will still need far more complicated sorting and dissecting to fully understand the mechanisms of drug effect. However this process works out, theory—whether it is behavioral or biological—has rarely predicted treatments in psychiatry or in the rest of medicine. We generally try things for the wrong reasons and learn better later.

What Did Thomas Detre Pioneer?

We obviously owe a debt to technology for our biological measures and their precision. But not the least important element of our progress in how we got here—which is what I am addressing—has been our capacity ethically to engage our patients in the commitment to treat as well as to search. This development since the mid-1950s has meant that the search could be in outpatient or in general hospital inpatient facilities, as well as in highly specialized settings or in communities. At Yale, within a year or two after his arrival in the mid-1950s, Thomas Detre led in this dramatic shift from sterile, remote, and isolated approaches to vigorous, combined therapeutic and research engagements with patients. Bench and bedside were joined in his ward. He turned his pioneering general hospital unit into a lively daily encounter with research as well as treatment. Dr. Detre has a unique talent for piercing taboos. He has a knack for sweeping away the nonessentials and for coming directly to the heart of the operative elements of a problem. Even social class taboos were demolished—professors, merchants, white- and blue-collar careerists, and homemakers; young and old; white, black, and yellow were mixed in his ward. Rituals and revered notions, such as the right to passivity in patienthood, were promptly abrogated. Rather than ritual, there was a collaboration of treatment and inquiry on the part of students, staff, and patients. All joined in his ward as he undertook research on drugs, on energy metabolism, on sleep, and—even then—on proprioception and schizophrenia. By 1964 the Intramural Programs of the National Institute of Mental Health (NIMH), although highly specialized, were also actively treating their research patients.

These developments have brought us to the present scene. Today, if a patient seeks a top subspecialist who knows more than others about what may be going wrong and what steps may best put it right, that subspecialist and therapist is likely to be busy doing research in a research-treatment setting rather than to be a practitioner who opines in leisure hours. At the least, today's expert practitioners are graduates of such settings. That change has gone unnoted. Yet I believe that it is centrally significant to both the vigor and rigor, as well as the relevance, of investigations in the clinical science of psychiatry at this juncture.

Research Strategies in Psychiatry—Are They New?

It is prudent not to be too contemptuous of the past. Table 10–2 simply reminds us that some of the topics we discuss at contemporary symposia have been discussed since the 1920s.

Our forebears were not dumber than we are. They were simply not as well equipped with information. Note for instance that by 1949, the link between life stress and bodily disease was an intellectually lively topic area. By that time, Schally and Guillimen were working on prohormones that are today central in psychiatric investigations. The brain, even then, was seen as an endocrine-generating and endocrine-responsive organ.

Further, the array of research strategies that were in use long before chlorpromazine made our lives a little easier does not differ much from those we use today. Longitudinal, follow-up, and epidemiological approaches to populations were a sturdy part of inquiry that, in the United States, was for a time largely forgotten by most psychiatrists in the explosive post–World War II growth of the field.

Energy hypotheses (now studied with high technology) had abounded since the late 19th century. Although some of the treatments of the past seem bizarre (e.g., endocrine surgeries, hyperbaric oxygen), we should not dismiss the underlying thought behind them. This was that mental disease is not a failure of will but somehow of the energy to deploy attention at will and to stay sturdily in touch with and responsive to the environment. In this sense, the biology of the psychiatric patient was suspected not to be efficient (and one recalls, if not Freud and neuraesthenia, then Janet and others essentially concerned with a weakened integrative capacity in psychopathology or with the Jacksonian notions of regression and levels of adaptation, and so on). Accordingly, sleep therapies and attempts to restore strength or to stimulate response were common.

Attempts experimentally to induce or relieve disorders were popular, as was the search for biological stigmata to identify disease. Certainly the notion of a psychotoxin that might directly trigger psychosis was an almost monomaniacal search ("schizophrenia is in the blood") that may now appear fre-

Table 10–2. Topics of some Association for Research in Nervous and Mental Disease (ARNMD) volumes, 1920–1953

Year	Topic
1920	Acute epidemic encephalitis (lethargic encephalitis)
1923	Heredity in nervous and mental disease
1925	Schizophrenia (dementia praecox)
1928	The vegetative nervous system
1930	Manic-depressive psychosis
1932	Localization of function in the cerebral cortex (frontal lobes, etc.)
1936	The pituitary gland
1938	The interrelationship of mind and body
1939	Hypothalamus and central levels of autonomic function
1944	Military neuropsychiatry
1947	The frontal lobes
1949	Life stress and bodily disease
1951	Psychiatric treatment
1952	Metabolic and toxic diseases of the nervous system
1953	Genetics and the inheritance of integrated neurological psychiatric patterns

netic. Of course, today we are searching for the endogenous chemical assassins that will destroy neurons or shift neuronal maturation and adaptive capacities. Investigators of the past studied the psychobiology of stress, arousal, and attention. Any clear focus on contemporary psychophysiology shows that we are still trying to disentangle these multiconnected mechanisms, and we still essentially use the measures of the polygraph to find links to disease. Investigators of the past used challenge paradigms (e.g., histamine injections and blood pressure response) and probes of invisible systems and subsystems within the brain and, as Herbert Meltzer and Gary Gudelsky have demonstrated (Chapter 9), we still do. Although they did not have pluckable genes in hand, they knew heredity was important. As did our predecessors, we all have theoretical models about how things might work—how the brain can operate to produce dissociations, dysregulation, and dysfunctions. Arvid Carlsson demonstrated for us his recent work on the multiple paths and reciprocal regulations by which arousal and cortical alerting and feedback may operate (Chapter 8). But I recall that some years ago, speculating on a model of brain operations and behavior, he noted that the heart was full of wires and juices just as the brain is. He asked the question that I can paraphrase as follows: Because disorders of the heart are disorders of rate and rhythm, why could these parameters not be relevant to neurons and behavior? In effect, he asked, should we not "think of functional pathology in terms of aberrations in the regulation of rates and rhythms," and, I would add, in the force and efficiency of the brain's output? I have noted elsewhere that we use analogies, metaphors, and, at best, models and paradigms to help us come to grips with the efficient and inefficient inherited designs of nature (Freedman 1982). These devices move us to empirical search. Our real problem has been to get the brain to operate in accord with the observations, intuitions, or paradigms that the imagination can propose. Of course, we also require that our concepts be tutored by scientifically revealed designs rather than revealed truth.

In sum, we subscribe to the view that what is biological about psychiatry is not determined by what we measure or how we treat but, rather, by how we think about what we study, discover, and do. In that vein, it is clear that the concept prevailing by 1950 was that external or internal signals will activate normally synchronized behavioral and bodily systems (we are still dissecting and identifying component systems involved with stress, panic, vigilance, and other traits). The response of these systems, it was thought, is mediated by various hormonal messenger systems linked to relevant neural subsystems. Concepts of that era would accommodate and predict many of the particular findings of modern neuroscience. What today's researchers may fail to note is the firmness of these regulatory and feedback notions and how such physiological thought actuated research. The view that mental disorder entailed some dysregulation in these systems was a congenial and central one to many of my teachers and to many of theirs (see Bernard 1927).

Finding a Brain That May Account for Behavior

Again, for historical purposes, one can see from where we have come (as epitomized in Table 10–3). Thus, I have called the pre-1940 picture that we had of the brain "Prometheus Bound." We had a picture of a brain that was able to produce "go" or "no-go" signals but, in terms of its known neuronal operations, could not say "maybe," "maybe if," or as Lewis Thomas once noted, even "what the hell!" Discoveries from the 1940s forward included moving the so-called "autonomous" nervous system firmly into the head—a development that really took about 40 years of work. I have (again telegraphically) noted these events in Table 10–3. Clearly that great internuncial (reticular) system lying between the large input and output tracks gave us a reactive brain capable of modulated and graded states of consciousness. These sensorimotor, reticular systems bring us physiologically into contact with both our internal and external environments and simultaneously enable variable awareness of them.

Table 10–4 sketches, perhaps too boldly, the elements of how this story unfolded. Around 1950, we knew about "long-loop" signaling such as adrenocorticotropic hormone (ACTH) signaling the adrenal, which released corti-

Table 10–3. Prometheus bound and unbound

Prometheus bound: brain pre-1940

1. Motor output and sensory input tracts and terminals known
2. Corticocentric: Motor and language centers, etc.
3. Action potential: Go or no-go; no "maybe" or "maybe if" or "what the hell!"

Prometheus "unbound": brain from the 1940s on

1. "Limbic lesions" and regressive sexual, eating, and grooming behavior; "autonomous" nervous system (1906); emotions moved from peripheral organs to the brain (1900–1950) with integrated autonomic and neurohormonal neural nets
2. Reticular systems—graded states: affect "readiness to go;" awareness; sensory input and outputs (see Starzl and Magoun 1951, etc.); subcortical and cortical links
3. Action potential: graded inhibitory or excitatory potentials found
4. Gamma efferents and feedback control on sensory input

Prior organization and state determine the fate of an input (intensity and direction of effects).

Brain is "built to behave." It is "set" to go ("corollary discharge"), to ignore or receive, simplify, shift, modulate, or "automate" signals, and to correct itself by response and feedback (action or thought).

sone, which in turn fed back to the brain. But the real problem was to get the brain to yield the secrets of any short-loop or interneuronal signaling (if any signaling other than electrical could be defined).

By 1950 we knew that drugs and hormones surely affected the brain. But the brain was a black box. We knew that glucose and oxygen were necessary for neuronal viability, but we could not (and as yet cannot) find Ralph

Table 10–4. Drugs, behavior, and "synaptology": 1945–1980s

1.	a)	Is there an intrinsic brain chemistry?
	b)	Is it affected by drugs and nutrients?
	c)	Is *behavior* concomitantly affected?
	d)	Do "classes" of drugs have different patterns of brain effects?
1946–1954		NE—1946 in brain; HT—1948 structure; and in 1954 found in brain (enzymes regulating synthesis and catabolism follow through the decade)
1955		Reserpine—NE and HT ↓ and waking behavior changes: not a drug but brain milieu of amines correlate with behavior
1957		MAOI—NE and HT ↑ (found first in periphery, then in brain)
1961		LSD—brain NE ↓ and HT ↑
1963		CPZ → change in brain DA synthesis and metabolites
1964	a)	DA depleted in Parkinson's disease
	b)	TCAs → change in NE metabolites (reuptake blockade)
	c)	Amine source cells visualized
1970s →		"Receptorology": signal dampening and amplification mechanisms, peptide receptors
1980s →		Anatomy of terminals and amine subsystems and colocalization with peptides further specified
2.		Does "experience" or environment affect brain structure and function?
	a)	"WIRES"—1952–1970s: Ocular stimulus deprivation and postnatal cellular change; synaptic "knobs," etc. Adolescence: Change in organization (sleep electrophysiology)
	b)	"JUICES"—1961–1963: Effects of stress on brain NE and HT In 1970s: stress changes 1) synaptic membrane (uptake) 2) receptors 3) biochemical regulations and equilibria of amines (brain slice) 4) cell potentials and ion flux: response bias without structural change; synaptic interneurons "learn" and "sensitize" or diminish effects of a signal
	c)	DEVELOPMENT and synaptic loss; increased corticolimbic O_2 use in critical age span: 2–14 years, etc.

Gerard's postulated "twisted molecule for every twisted thought." There indeed may be tangled hippocampal neurons and disarray in the modular arrangements Goldman-Rakic addresses. Seymour Kety had to conclude, after studying humans' brain oxygen consumption, that it took no special energy to think a normal or crazy thought. We have, of course, since moved on to regional blood flow studies and PET scan measures that would revise such dicta, but my point is simply that the important questions preceded the technology.

What really changed things for us were the psychotropic drugs. We have four fundamental questions to remind today's scientists about. In 1950 we knew that brain function was changed by drugs and nutrients, but whether there indeed was an intrinsic brain chemistry or, second, whether it was specifically affected by drugs and nutrients was not established on the basis of direct evidence. Third, we did not know whether behavior was coupled to any such chemistry. Fourth, we did not know whether "classes" of psychotropic drugs would have different patterns of brain effects. Between 1955 and 1964, those questions were answered in the affirmative.

For me, the key observation was that reserpine depleted the amines and that the waking behavior of the animal concomitantly changed. With further study, it became clear that it was not the drug but the altered milieu of brain amines that correlated with the changed behavior. That is what sparked the excitement about linking aminergic systems to behaviors. When MAOIs were accidentally discovered to raise the brain levels of amines and LSD was found to have a different set of effects on amines, it was becoming clear that classes of drugs indeed had special effects.

My colleagues and I have been tracking LSD since my first observation of its effects on brain serotonin in 1960 (Freedman 1961) and most recently have tracked these down to the 5-HT_2 receptors (Buckholtz et al. 1988; Freedman 1986). Not only do specific 5-HT_2 receptor blockers (such as ritanserin) block LSD behavioral effects in animals (and by rumor from the streets, those in humans), but the remarkable tolerance after daily doses of LSD in rats and humans is accompanied by a parallel down-regulation of the 5-HT_2 receptor (Buckholtz et al. 1990). Yet we still do not know how that remarkable LSD induced "TV show in the head" gets turned on. We knew in the mid-1950s that the reticular system of an LSD treated animal was more readily activated by any kind of environmental input (just as Aghajanian's elegant recent studies show inputs to the locus coeruleus [Aghajanian 1980] or the facial motor nucleus [McCall and Aghajanian 1980] to be facilitated by hallucinogens). Unlike amphetamine, hallucinogens in the cerveau isolé did not generate signals but, rather, enhanced reactivity to them. We know clearly that the effects of the drug and their intensity are serotonin linked and that if we deplete serotonin even by 10% we can increase the sensitivity of response to LSD fourfold—a dose four times less than the threshold behavioral dose is effective (Freedman 1986). So we know the amines modulate events after LSD meets the receptor, and we also know they do not directly "turn on" these remarkable trips within the head.

By 1963, Carlsson had shown us what would happen following chlorpromazine administration by measuring amine metabolites and synthesis; and by 1964, Axelrod had essentially demonstrated reuptake blockade. The basic anatomy of the chemical brain was illumined by the Swedish fluorescent microscopy that identified the amine source cells in the base of the brain. Further, we knew that brain chemistry and membrane functions were changed by environmental demands as well as by drugs. We have since moved on to the age of "receptorology" and recently to the cortical (often modular) anatomy of the distal nerve terminals of noradrenergic and serotonergic source cells.

Clinical Research in Psychiatry and Biomedicine: Similar Strategies

With all this promise and more, we still lack diagnostically specific biological definitions of prior state in untreated patients; we lack knowledge of etiopathology. We have had some good glimpses of pathophysiology but many more definitions of compensatory therapeutic mechanisms. Although we often forget it, this is true for 80% of internal medicine disorders. We in psychiatry—addicted to self-examination, if not self-flagellation—should clearly appreciate that the strategies we employ are exactly those used for most diseases. (In fact, in some areas involving psychometrics and rating scales, as well as multisite clinical trials and epidemiology, we have led the way.) Any medical text reveals that, even with today's fancy measures, processes related to causes of disorders and certainly to their phases or to disease outcome still require search.

This truth is well elucidated by grasping the situation for diabetes (as Table 10–5 shows). Only since the 1970s have the different physiological and molecular mechanisms that lead to elevated blood sugar begun to be identified. A distinct familial clustering was long evident but only recently clarified for Type I or Type II diabetes.

I highlight the fact that biomedical research now focuses on disease subtypes, on receptor fluctuations, and on autoimmune mechanisms affecting receptors or substrate. Further, note that the parameters described for the study of diabetes are precisely those used in studying schizophrenia. Thus, we look at symptomatically silent prior states, immediate states, and long-term events. Each phase requires different explanations, interventions, and research strategies. Recall that we had insulin for 50 to 60 years before this newer and still developing knowledge became apparent. Psychiatrically useful drugs have been around for half that time and our target organ is far more complicated.

Thus, I believe we must calibrate our expectations as we piece together the various elements of the story and where they fit. In all of medicine, we usually have partial stories. So it is and so it has been. It is a long trajectory

Table 10–5. Diabetes as a prototype

1920–1970: Diabetes, or high blood sugar

1. Insulin—secreted by pancreas—lowers high blood sugar— "insulin deficency" disease
2. Control of environment: diet, exercise, stress, or infection—all change need for injected insulin
3. Significant familial clustering

1970–present: Diabetes, a "final common path"

1. Insulin levels can be the same—*effect* on cells different
   a) "defective" insulin molecule—does not "fit" cell receptor
   b) overweight person with diabetes loses weight: same "natural level of insulin is now more effective; more cell receptors "grow"
2. "High insulin" levels are ineffective: subsensitive cell receptors ("damaged" by antibodies; antibody to insulin as well)
3. Adult onset: high monozygotic (MZ) twin concordance
4. Juvenile onset: less gene "control." Genetic "vulnerability" depends on environment interactions (e.g., viral infection for *expression* of diabetes)
5. Research on the "vulnerable" and "at-risk"
   a) in animal models and twins
   b) MZ twins—*not always concordant* (postnatal rearrangement of genes in immune systems can occur)
6. Current focus on *symptomatically silent long prodromal phases*—search for "markers" and molecular and environmental determinants of disease expression
7. Etiopathology of *late phase* neurologic and cardiovascular effects a research focus
8. Risk of *mildly* elevated blood sugar levels now being defined

For perspective

Asymptomatic prior states, immediate state, and long-term events require different explanations, interventions, and research strategies.

We knew of insulin for 50–60 years before newer knowledge, and of specific drugs in psychiatry for about half that time. Our specialty's "organ" is more complicated.

involving oscillations between clinical observations, basic science, accidents of nature, and serendipity. We cannot require the "eureka!"-like leap from the unknown to the definitive final answer (Freedman 1987). We cannot even expect such leaps from our current wonder drugs (though we surely would warmly accept them). If we look at our essential strategy in biological psychiatry, I can stand by the credo that Herbert Meltzer and I formulated in 1975 (Freedman 1975). It was that, for the study of disease, we examine the entire span of clinical phenomena from prodrome to outcome, from primary mechanism to secondary and tertiary consequence. We do so to see what the clinical parameters reveal about the significance of biological measures and to see how biological measures relate to some aspect of the clinical parameters.

We are, for example, today measuring 3-methoxy-4-hydroxyphenylglycol (MHPG) and many other substances and metabolites. Notice that we can rarely speak of an unequivocally abnormal value. Rather, what we find are correlations with components of disease. The same absolute value may be normal or reflect the operation of an ongoing compensatory process (Freedman 1987). We have learned that amines may be "taxed" when other adaptive systems are impaired; equally, amines or their regulation may be intrinsically faulty in the presence of psychologically intact coping systems.

The problem before chlorpromazine was that vast uncharted territories separated the few measurable substances from which we tried to deduce what is going on. The distances have diminished but are still aggravatingly present. Absent good fortune, a painstakingly detailed physiology to link relevant processes is still required with our modern tools.

Thus, I have referred to the "upstairs/downstairs" problem—we still do not know how many peripheral measures are linked to the central nervous system. In brief, there indeed are phases in the work of science and only occasionally do we have moments of clarity. More frequently we are faced with bafflement. About 50% of the DNA is devoted to the brain, as we have heard. And there is much yet to be mined (Freedman 1987). More commonly, elegant solutions yield to a new phase of puzzling details seeking meaning, and there are expectable periods of time when we move from clarity to a kind of inattention to why we are measuring what we are as flurries of details are assembled.

Psychiatry, Medicine, "Clinical Neuroscience," and Biology

It is an old rule in science that every question answered generates new ones. Fundamentally, then, we should realize that psychiatry and medicine are part of the life sciences. We should come to understand the range of mechanisms, both endogenous and exogenous to nuclear DNA, by which apparently similar phenotypes can be observed. There are many, perhaps too many, such pathways. The intrinsic genetic design, DNA, "expects" an environment. Its very structure and blank areas provide for order, error, mutability, and chance from conception and throughout the life cycle. This instructional and expressive design entails the ceaseless interplay of the "seeking and the sought" through molecular, cellular, and behavioral communications. In embryology we see a nerve seeking the target that signals an "interest," and in the human we see a search for sources signaling emotional nurturance (Freedman 1987).

At bottom, then, we are built to behave. We have a brain that is also a dynamic system built to behave and, like DNA, also one with "blank spaces." It constructs a history as it develops so that the organism can acquire information, memory, and attitudes—all of which help in coping with the environment in an expectable fashion. We know that learned familial rules, signals, and personal memories partially structure our interpretation of stress and that the "psychosocial" is a biologically rooted part of the regulatory operations that we must study (Freedman 1987). A century ago, Pascal asked, If nature is first habit, why is not habit second nature? This disposition for entraining during the age period when societies "capture the mind of the child" (Freedman 1978) has been captured in PET scans of the early phases of life. Thus, Chugani and colleagues (1987) showed that cortical but not cerebellar areas show a huge excess of energy utilization between the ages of two or three and puberty as a great deal of the person's ultimate destiny is being established.

If these shadowy perceptions of "the way things are" at all resemble the intrinsic phenomena with which we deal, we might search for some coordinates with which to assess our progress. I believe there are some overblown fantasies to the effect that clinical judgment in medicine can soon be dispensed with. Without elaborating, I believe that no fictional all-embracing "clinical neuroscience" will wholly replace psychiatry. I do not believe that an image of the lions and the lambs of neurology, neurosurgery, psychiatry, and molecular neuroscience peacefully grazing the pastures of the brain together is a useful picture of what we must inevitably confront. For it is the different diseases these clinical sciences are charged to comprehend that pose their own agenda of problems and needed interventions and, hence, of questions to be posed of the indifferent complexities of nature. Further, this is not an issue of entrenched guilds; diseases per se force specialized attention. Ultimately, if one truly understands the inexorability of chance and variation intrinsic to biology, there is no escape in dealing with disease variations and the varyingly different domains that account for them. All biomedicine might yield the underlying notion that we can ultimately control all outcome in the name of therapy or prevention. Biological systems usually do not work that way. Nature does not need the illusions of omnipotence that console us

(Freedman 1987). In brief, it is a truism that if we abolished psychiatry, a race of "psychiatroids" would have to evolve to replace us.

Yet patience, clever experimental design, and the sophisticated pursuit of probes that indicate intrinsic design can bring us far and will take us farther. Although we all latently wish for a truly exact meeting of mind and matter, we should, rather, be excited by understanding how a fruitful meeting might in reality be effected. That meeting was something our forebears in psychiatry yearned for. They were fascinated with the mind-body problem—how it is that living matter can accommodate the links not only of bodily function but of subjective experience and of behavioral as well as symptomatic patterns of performance (Freedman 1978). They, like us, pondered models for this linkage and generated research strategies identical to ours.

They, like us, believed that if we understand neurobehavioral operations, precognitive performances in their simpler forms, we might gain a better grasp of operations necessary for more complex human behaviors. Kandel, in his studies of snails, for example, may have given us an ego for the escargot. Yet we have seen in mammalian systems that processes—sensitization, habituation, long-term potentials, and more—will be key in understanding how environmental signals are amplified or dampened and how behaviors become entrenched or too readily elicited. There is, thus, profound clinically relevant work yet to be done in psychobiology, but one can see in the approaches of all contributors to this volume that we are learning to ask both disease *and* psychobiological questions in an excitingly productive way. We are asking both microsystem and macrosystem questions to understand the physiology of that remarkable organ of adaptation.

Knowing the depth, extent, and reach of research currently under way, we can envisage yet another lively meeting of mind and matter that will overcome some of the distances that prevail in our current ignorance. We can feel grateful to those who contributed to this book and especially to the person who is the inspiration for it. Together, they have given us the privilege of envisioning a long-sought future.

References

Aghajanian GK: Mescaline and LSD facilitate the activation of locus coeruleus neurons by peripheral stimuli. Brain Res 186:492–498, 1980

Baxter LR, Schwartz JM, Mazziotta JC, et al: Local cerebral glucose metabolic rates in obsessive-compulsive disorder. Arch Gen Psychiatry 44:211–218, 1987

Bernard C: Experimental Medicine. Translated by Greene HC. New York, Macmillan, 1927

Buckholtz NS, Zhou D, Freedman DX, et al: Serotonin-2 agonist administration down-regulates rat brain serotonin-2 receptors. Life Sci 42:2439–2445, 1988

Buckholtz NS, Zhou D, Freedman DX, et al: Lysergic acid diethylamide (LSD) administration selectively downregulates serotonin$_2$ receptors in rat brain. Neuropsychopharmacology 3:137–148, 1990

Chugani HT, Phelps ME, Mazziotta JC: Positron emission tomography study of human brain function development. Ann Neurol 22:487–497, 1987

Frank E, Kupfer DJ, Perel JM, et al: Three-year outcomes for maintenance therapies in recurrent depression. Arch Gen Psychiatry 47:1093–1099, 1990

Freedman DX: Effects of LSD-25 on brain serotonin. J Pharmacol Exp Ther 1:653–658, 1961

Freedman DX (ed): The biology of the major psychoses: A comparative analysis. New York, Raven, 1975

Freedman DX: From mind to brain: new emphasis on psychiatric education. Yale J Biol Med 51:117–131, 1978

Freedman DX: Presidential address: science in the service of the ill. Am J Psychiatry 139(9):1087–1095, 1982

Freedman DX: Hallucinogenic drug research—if so, so what?: symposium summary and commentary. Pharmacol Biochem Behav 24:407–415, 1986

Freedman DX: Strategies for research in biological psychiatry, in Psychopharmacology, the third generation of progress: The emergence of molecular biology and biological psychiatry. Edited by Meltzer HY. New York, Raven, 1987

McCall RB, Aghajanian GK: Hallucinogens potentiate responses to serotonin and norepinephrine in the facial motor nucleus. Life Sci 26:1149–1156, 1980

Redlich FC, Freedman DX: The Theory and Practice of Psychiatry. New York, Basic Books, 1966

Starzl TE, Magoun HW: Organization of the diffuse thalamic projection system. J Neurophysiol 14:461–477, 1951

Valenstein ES: Great and Desperate Cures. New York, Basic Books, 1986

Index

Note: Page numbers printed in boldface type refer to tables or figures.